The behavior of structures composed of composite materials

MECHANICS OF STRUCTURAL SYSTEMS
Editors: J.S. Przemieniecki and G.Æ. Oravas

L. Fryba, Vibration of solids and structures under moving loads. 1973. ISBN 90-01-32420-7

K. Marguerre and H. Wölfel, Mechanics of vibrations. 1979.
ISBN 90-286-0086-6

E.B. Magrab, Vibrations of elastic structural members. 1979.
ISBN 90-286-0207-0

R.T. Haftka and M.P. Kamat, Elements of structural optimization. 1985.
ISBN 90-247-2950-5(hardbound) ISBN 90-247-3062-7(paperback)

J.R. Vinson and R.L. Sierakowski, The behavior of structures composed of composite materials. 1986.
ISBN 90-247-3125-9

The behavior of structures composed of composite materials

J.R. Vinson

Department of Mechanical and Aerospace Engineering
University of Delaware
USA

R.L. Sierakowski

Department of Civil Engineering
Ohio State University
USA

1986 **MARTINUS NIJHOFF PUBLISHERS**
a member of the KLUWER ACADEMIC PUBLISHERS GROUP
DORDRECHT / BOSTON / LANCASTER

IV

Distributors

for the United States and Canada: Kluwer Academic Publishers, 190 Old Derby Street, Hingham, MA 02043, USA
for the UK and Ireland: Kluwer Academic Publishers, MTP Press Limited, Falcon House, Queen Square, Lancaster LA1 1RN, UK
for all other countries: Kluwer Academic Publishers Group, Distribution Center, P.O. Box 322, 3300 AH Dordrecht, The Netherlands

Library of Congress Cataloging in Publication Data

Vinson, Jack R., 1929–
    The behavior of structures composed of composite
materials.

    (Monographs and textbooks on mechanics of solids and
fluids. Mechanics of structural systems ; 5)
    Includes bibliographies.
    1. Composite materials. 2. Composite construction.
I. Sierakowski, R. L.   II. Title.   III. Series.
TA418.9.C6V55 1985   620.1'18   85–3064
ISBN 90–247–3125–9

ISBN 90-247-3125-9 (this volume)
ISBN 90-247-2735-9 (series)

Copyright

PRINTED IN THE NETHERLANDS

This book is dedicated to our wives Midge and Nina for their patience, encouragement and inspiration.

For Midge Vinson, appreciation is expressed for providing the love, care, and environment for keeping her husband productive and well. This text could not have been completed without her.

For Nina Sierakowski, appreciation is expressed for patience and understanding during this undertaking as well as for caring for Steven and Sandra and teaching them the art of periodic silence.

Preface

While currently available texts dealing with the subject of high performance composite materials touch upon a spectra of topics such as mechanical metallurgy, physical metallurgy, micromechanics and macromechanics of such systems, it is the specific purpose of this text to examine elements of the mechanics of structural components composed of composite materials. This text is intended for use in training engineers in this new technology and rational thought processes necessary to develop a better understanding of the behavior of such material systems for use as structural components. The concepts are further exploited in terms of the structural format and development to which the book is dedicated. To this end the development progresses systematically by first introducing the notion and concepts of what these new material classes are, the fabrication processes involved and their unique features relative to conventional monolithic materials. Such introductory remarks, while far to short in texts of this type, appear necessary as a precursor for engineers to develop a better understanding for design purposes of both the threshold limits to which the properties of such systems can be pushed as well as the practical limitations on their manufacture.

Following these introductory remarks, an in-depth discussion of the important differences between composites and conventional monolithic material types is discussed in terms of developing the concepts associated with directional material properties. That is, the ideas of anisotropic elasticity for initially homogeneous bodies in the phenomenological sense are described and presented. The use of such analytical tools is then presented through exemplification of selected problems for a number of classical type problems of various geometric shapes including plane stress, plane strain and the bending of a simply supported beam.

These ideas are carried forward and developed for continuous fiber composites in Chapter Two which discusses both single ply laminae and multi-ply laminate theory. This is then followed by a series of chapters, each of which deals with a functional aspect of structural design in which the basic building blocks of a structural system are made. That is, plates and panels; beams, columns and rods; and cylindrical and spherical shells are each discussed within the framework of their potential use in a functional environment. Thus the traditional topics of conventional monolithic (isotropic) material structural elements such as structures subjected to static loads, thermal and other environmental loads, structural instability and vibratory response are included along with chapters on energy methods and failure theories of composite materials.

Energy methods have been included to present a tool for solving difficult problems of various types encountered in practice. Indeed, in many instances closed form solutions are not possible and approximate solutions must be sought. Energy methods thus provide both an alternative for the formulation of such problems plus a means of generating approximate solutions.

The chapter on failure theories is a generic presentation in the sense that any and/or all of the above structural components consisting of various multi-ply construction can fail when subjected to a sufficiently large loading combination. It is emphasized that the failure of composites is a complicated, changing issue because of the diverse ways in which such structural systems can fail due both to the geometric ply arrangement of the components, complicated load paths, and the diversity of failure mechanisms which can be activated. Therefore, this chapter should serve in a global sense at best as a guide to the prediction of structural integrity, while more common and acceptable phenomenological failure theories are being developed.

Finally, a chapter on joining is included to discuss to some detail the two methods by which composite material structural components can be joined: namely, adhesive bonding and mechanical fastening. Again, the material presented is an introduction to the subject which is rapidly changing and developing.

At the end of each chapter are several problems, characteristic of the material covered which can be used. Some answers are given in an appendix.

Knowing that nothing is perfect, the authors welcome any notification of errors and ambiguities, and if addresses are provided, authors will forward errata sheets periodically.

Appreciation is hereby expressed to the many students at the University of Delaware, University of Florida, Ohio State University, The Ballistics Research Laboratory, and the Argentine Air Force who have helped directly or indirectly in refining, improving and correcting the text, as well as working various problems and examples. In addition appreciation is expressed to Dr. W.J. Renton, Vought Corporation, who has used portions of the text at the University of Texas-Arlington, and made suggestions and corrections.

Jack R. Vinson
Robert L. Sierakowski

Contents

1. Introduction To Composite Materials

1.1. General History

The combination of materials to form a new material system with enhanced material properties is a well documented historical fact. For example, the ancient Jewish workers during their tenure under the Pharaohs used chopped straws in bricks as a means of enhancing their structural integrity. The Japanese Samurai warriors were known to use laminated metals in the forging of their swords to obtain desirable material properties. Even certain artisans from the Mediterranean and Far East used a form of composite technology in molding art works which were fabricated by layering cut paper in various sizes for producing desired shapes and contours.

In order to introduce the reader to the subject matter of new high-performance composite materials it is necessary to begin by defining precisely what constitutes such a class of materials. Furthermore, one must also define the level or scale of material characterization to adequately describe such systems for discussion. This is done with the understanding that any definition and classification scheme introduced is arbitrary and within the province of the writer.

For introductory purposes, many workers in the field of composites use a somewhat loose description for defining a composite material as simply being the combination of two or more materials formed to obtain some useful new material or specific material property. In some cases the addition of the words microscopic and macroscopic are added to describe the level of material characterization.

The definition posed above is to a large extent broad-based, in that it encompasses any number of material systems for which different levels of characterization must be used to specify the system and for which different analytical tools may be necessary for modeling purposes. As a simplistic example of the definition used above we can consider a beam consisting of clad copper and titanium material elements used in a switching strip. Such a composite system can be considered at the macroscopic level as providing enhanced temperature-dependent material behavior due to the mismatch in thermal coefficients of expansion between the copper and titanium metallic elements. This material system, while consisting of two dissimilar materials and falling within the realm of satisfying the definition of a composite material would not be acceptable as being representative of modern definitions of composites for current applications in the aerospace, automotive and other technical areas. A representative bibliography of books and periodicals dealing with composites has been included at the end of this chapter.

1.2. Composite Definition and Classification

In order that we might agree at the outset on a suitable modern day definition for advanced composite materials we introduce a structural classification schedule according to the use of the following typical constituent elements.

STRUCTURAL LEVELS
(I) BASIC/ELEMENTAL Single molecules, crystal cells
(II) MICROSTRUCTURAL Crystals, Phases, Compounds
(III) MACROSTRUCTURAL Matrices, particles, Fibers

Of the structural types cited above, Type (III), or the Macrostructural type is the most important for our further discussion. Continuing with this, we next consider a further classification schedule within the structural framework adopted. We are thus led to the following tabular schedule defined in terms of the fundamental classes of composites.

TYPES OF COMPOSITES

* Fiber
  Composed of Continuous or Chopped Fibers
* Particulate
  Composed of Particles
* Laminar
  Composed of Layers or Lamina Constituents
* Flake
  Composed of Flat Flakes

* Filled/Skeletal
  Continuous Skeletal Matrix Filled by Second Material

Schematics of each of these composite types have been included in Figure 1.1 below.

Within the framework of the above classification schedule, one of the most important of the composite material types cited from an applications point of view is the filamentary type, that is, those material systems consisting of selected fiber macroconstituents. Such material systems have the desirable properties of having high strength and/or stiffness as well as being light in weight. These are the materials with which we are principally concerned. In order to identify these material types one can examine the Periodic Table of Elements in Figure 1.2 to select those lightweight components from which filament elements can be formed.

1.3. Filamentary Type Composites

As a point of reference we now introduce the following definition for a fibrous composite material. "Composites are considered as combinations of material elements which differ in composition or form on a macroscopic level with respect to each other. The individual fibrous constituent elements can be man-made, are generally insoluble, retain their identities within the composite, and may be continuous or discontinuous."

A chart indicating the classification and types of filamentary type systems existing in practice is shown in Fig. 1.3.

In order to obtain a feeling for the representative types of fibrous composites we present a summary of some of the more common fibers Fig. 1.4 and matrix materials Fig. 1.5 used in practice. This listing appears without attention focused on any of the specific types of fiber composite systems that are nonmetal matrix (NMMC) and metal matrix (MMC) composites, as used in practice.

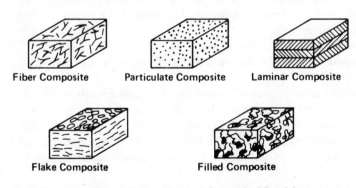

Fiber Composite   Particulate Composite   Laminar Composite

Flake Composite   Filled Composite

Figure 1.1. Illustration of Various Types of Composite Materials.

4

PERIODIC CHART OF THE ATOMS

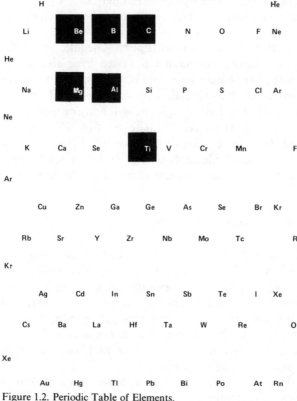

Figure 1.2. Periodic Table of Elements.

Of the reinforcements, Boron, Glass, Graphite, and Kevlar[†] have received the most attention in the literature.

The reader is also apprised of special codes to denote graded performance ranking of classes of composites. For example Gr_m refers to High Modulus Graphite.

The reinforcements and matrix materials cited are then combined to form composite structural configurations, as shown in the accompanying diagram, Fig. 1.6.

As noted, the combination of matrix and fibrous elements results in a composite structure that may have different structural forms by virtue of the forming and/or processing method used. We will say more later about the forming and processing technology used to produce the composite. The specific structural forms obtained can then be further synthe-

[†] Kevlar: Trademark of E.I. DuPont Co.

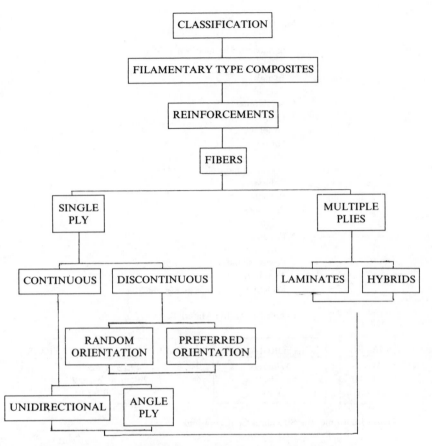

Figure 1.3. Chart of Various Types of Filamentary Composite Materials.

REINFORCED AND HIGH-PERFORMANCE COMPOSITES

Reinforcements

Alumina	$-Al_2O_3$
Aluminum	$-Al$
* Boron	$-B$
Boron Nitride	$-BN$
Beryllium	$-Be$
* Glass	$-Gl$
* Graphite	$-Gr$
* Aramid (Kevlar[†])	$-Kv$
Silicon Carbide	$-SiC$
Silicon Nitride	$-Si_3N_4$
Titanium	$-Ti$
Tungsten	$-W$

* Principal High-Performance Reinforcements

Figure 1.4. Reinforcements.

MATRIX MATERIALS

(Thermosets, Thermoplastics, Metals)

        Acetals
        Acrylics
*   Aluminum
*   Epoxy
        Graphite
        Nickel
        Nylon
        Polybenzimidazoles
        Polyesters
        Polyethylenes
*   Phenolics
        Polypropylenes
*   Polyimides
        Polyurethanes
        Phenyl silanes
        Titanium
*   Principal high-performance matrices

Figure 1.5. Matrix Materials.

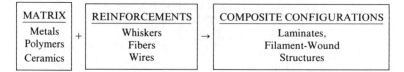

Figure 1.6. Composite Structural Configurations.

sized through fabrication and forming procedures in order to satisfy the required design and/or functional applications.

Within the framework of typical reinforcements, mention is made of whiskers, fibers and wires. We may identify such reinforcements by their diameter (average diameter in the case of fiber bundles) and consider the following Fig. 1.7 as a useful reference for such purposes.

1.4. Composite Manufacture, Fabrication and Processing

As previously indicated, the synthesis of matrix and filament to form a composite structural configuration is achieved through appropriate forming processes. A selected list of these fabrication processes as related to forming both resin matrix and metal matrix composites follows, Fig. 1.8, with accompanying remarks on a number of the cited processes. It should be noted that each of these processes has certain inherent advantages

Reinforcement Sizes

Type	Diameter
Whisker	$< 0.001''$
Fiber	$0.001''-0.032''$
Wire	$0.032''-0.25''$
Rod	$0.25''-2.00''$
Bar	$> 2.00''$

Figure 1.7. Codification According to Size of Reinforcement.

associated with production, design, and economic circumstances. Also, in order to extend the scope of remarks related to composite material fabrication and processing, the presentation has not been restricted to the high-performance group of materials and particularly those of the continuous fiber reinforcement type. Thus, a somewhat broader survey of composite processes has been presented with tabulation of these processes listed in the accompanying chart, Fig. 1.8 for easy reference and use by the reader.

It should be noted that to a large extent composite manufacture is primarily influenced by the matrix. For example, organic matrix composites are generally laminated, compression-molded or extruded, with the specific method selected being dependent on a number of specific requirements, principal among which is cost-effectiveness. On the other

FABRICATION/ PROCESSING TECHNIQUE	(CNMMC) CONTINUOUS NON METAL MATRIX COMPOSITE	(CMMC) CONTINUOUS METAL MATRIX COMPOSITE	(SFC) SHORT FIBER COMPOSITE
Hand Layup	×		×
Vacuum Bag/Autoclave	×		×
Matched Die Molding	×	×	×
Filament Winding	×	×	
Pressure & Roll Bonding		×	
Plasma Spraying Techniques		×	×
Powder Metallurgy		×	×
Liquid Infiltration	×	×	×
Coextrusion		×	
Controlled Solidification		×	
Rotational Molding		×	×
Pultrusion		×	
Injection Molding		×	×
Centrifugal Casting		×	×
Sheet Molding Compound			×
Pneumatic Impaction		×	

Figure 1.8. Chart of Various Fabrication Processes.

hand metal matrix composites are generally laminated, cast, rolled, or extruded. Finally, chopped and discontinuous fiber composites are generally produced by matched die molding, rotational molding, injection molding, or centrifugal casting.

1.4.1. Powder Metallurgy

In this technique whiskers or chopped fibers are selected as reinforcing materials. The aligned or unaligned materials are then mixed with an appropriate powder matrix introduced into a die, either hot or cold pressed under high pressure, compacted, and in some instances sintered.

This technique suffers in that damage to the fibers can result from the pressing operation if the powder size is greater than the mean diameter of the fibers. Also, maintaining fiber alignment and minimizing porosity are also effects to be considered. Some systems fabricated using this technique are aluminum/stainless steel, Al-Ti alloy/Molybdenum and aluminum/tungsten.

1.4.2. Pressure and Roll Bonding

In this technique appropriate layers of sheets of the matrix material and aligned fibers are placed alternately atop one another, the combination heated to high temperature, and then either pressed together or rolled to form the composite system. For this technique to be functional both the sheet stock and filaments used in the processing must be extremely clean.

One of the inherent problems in using this technique is that reactions at the fiber-matrix interfaces may occur during the long times associated with the pressing operation.

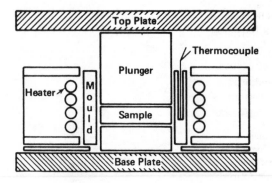

Figure 1.9. Powder Metallurgy Method.

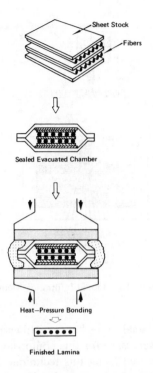

Figure 1.10. Pressure Bonding Method.

Some systems made using this technique are stainless steel/aluminium and tungsten/aluminum composites.

1.4.3. Liquid Infiltration

In general, two techniques are commonly used for this type of processing. These techniques are (a) the molten matrix can be either introduced into a mold assembly containing the aligned and packed fibers, or (b) alternatively the fibers can be introduced into the molten matrix. When using the first technique it is common practice to introduce the molten matrix into the system using either gravity or vacuum suction. The specific technique used is directly dependent upon the particular matrix/fiber system selected in that for effective wetting systems the matrix can be infiltrated and solidification allowed to occur, while in a system with less effective wetting action vacuum drawing is necessary, with the matrix drawn against the aligned fibers. In still other systems it may be necessary to coat the fibers initially in order to enhance the overall wetting action.

A major problem with using this technique is the inherent reaction

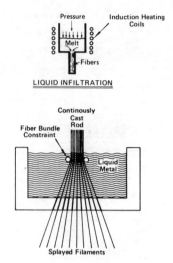

Figure 1.11. Liquid Infiltration Method.

produced in the interfacial zone between fiber and matrix due to the thermal environment necessary for this composite process to occur.

Some systems fabricated using this technique are steel-epoxy, boron-epoxy, glass-epoxy, copper-tungsten, molybdenum-copper and steel-aluminum.

1.4.4. Coextrusion

In this method combined elements of constituents are passed through appropriate dies under high pressure producing an extrusion consisting of the reinforcing phase and matrix material.

Due to the high pressures involved in the use of this technique it appears to be suited for use with soft matrix type materials. This fabrication process does possess an inherent advantage over others in that interfacial problems between the matrix and fiber are minimized. Also, since work hardening of the matrix is an inherent part of the forming process, it contributes to the strength of the composite. Such systems as NiCr alloys reinforced by sapphire whiskers have been formed in this manner.

1.4.5. Plating/Spraying Techniques

Both plasma spraying and electrodeposition represent some of the more familiar methods associated with the general plating/spraying technique. In plasma spraying a layer of fibers is wound on a rotating

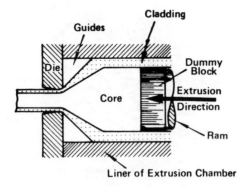

Figure 1.12. Coextrusion.

mandrel with the matrix metal deposited on the fibers by plasma spraying the matrix material from a powder feed directly onto the mandrel. A second layer of fibers is put on and the operation repeated until the desired number of layers have been obtained. The composite is then removed from the mandrel and can be reprocessed by hot pressing to remove any porosity.

The electrodeposition technique utilizes a conducting mandrel as a cathode in conjunction with a consumable anode of the matrix material in an electrolytic bath. A layer of continuous filament material is wound on the mandrel and then a layer of matrix material deposited on the wound filaments, with the amount of material deposited controlled by the rate of deposition of the matrix material and required filament spacing.

Some examples of materials fabricated using these techniques are steel-aluminum, boron-aluminum, and boron-titanium.

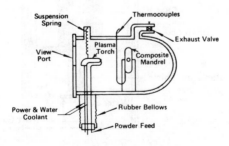

Figure 1.13. Plasma Spray Process.

12

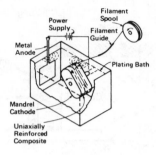

Figure 1.14. Electrodeposition.

1.4.6. Controlled Solidification

In this technique a cylindrical sample is drawn at a controlled rate through the eutectic point producing a unidirectionally solidified sample. The main advantage of this system is that the reinforcing phase does not depend upon a man-made procedure for filament placement resulting in a mechanically and chemically compatible system. Some of the systems formed using this technique are Copper-Aluminum ($CuAl_2$) and Aluminum-Nickel (Al_3Ni).

1.4.7. Hand Lay-Up

The preparation of controlled organic matrix composite samples for laboratory testing and data collection is usually accomplished by using

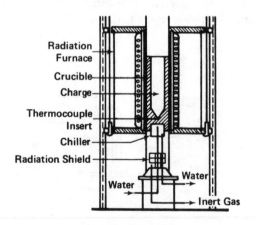

Figure 1.15. Controlled Solidification.

hand lay-up techniques. For industrial applications, machine lay-up is used in order to obtain low void content and cost-effective composites. In the case of filamentary composites, prepreg tapes are generally employed, with assembly consisting of placement of tapes according to predetermined design criteria. That is, the required ply orientation and stacking sequences are tailored by design requirements during the fabrication process. The specimen fabrication is generally assembled within the confines of a clean room using a template or tool plate as a reference surface. The fabrication process consists of selected a proper number of ply-on-ply sequences, along with appropriate peel and bleeder plies, to insure easy removal of the prepared specimens and to ensure absorption of excess resin during the cure cycle. The system is then processed by using either a vacuum bagging or autoclave cure technique to be described. A schematic of the hand lay-up method is shown in Figure 1.16.

1.4.8. Spray-Up (Open Molded)

A combination chopper and spray gun is used to simultaneously deposit fiber-glass chopped roving and resin onto the mold. Immediately after the combination is applied to the mold surface, it is rolled out by hand to remove air, lay down the fibers, cause the resin and glass combination to become densified and assure intimate contact with the mold surface. Cure of the component is then accomplished by moderate heat application and the catalyst system.

Applications include commercial refrigeration components, cabinets for reach-in units, coolers, and display cases, and laboratory fume hoods.

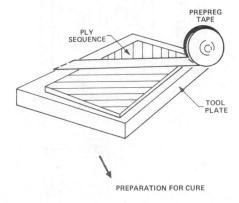

Figure 1.16. Hand Lay-Up or Machine Lay-Up Method.

14

1.4.9. Filament Winding

Filament winding was perhaps one of the earliest of the fabrication processes used for producing continuous filament composites. In the laboratory this process can be used to form controlled composites in two optional formats. In the first, the filaments are preimpregnated by passing them through a suitable matrix, while the second option requires winding the fibers onto a mandrel and impregnating the total filament assembly (Figure 1.17). Single filaments/strands or rings can be used in the first approach. Thus, the system can be used to form both organic and metal matrix composites. When a sufficient number of layers of fibers have been appropriately laid down, the impregnated windings are cured. Alternatively, if the second process is used, the mandrel is impregnated and allowed to furnace cool following a programmed procedure.

Filament winding uses continuous reinforcement to achieve efficient utilization of fiber strength. Roving or single strands are fed through a bath of resin and wound onto a suitably designed mandrel. Preimpregnated roving can also be used. Special lathes lay down continuous glass fibers in a predetermined patter to give maximum strength in the directions required. When the layers have been applied, the wound mandrel is cured at room temperature or in an oven.

Typical applications are chemical tanks and pressure bottles.

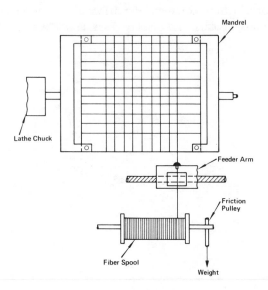

Figure 1.17. Filament Winding Method.

1.4.10. Vacuum Bagging / Autoclave Cure

Specimens prepared by either hand lay-ups or machine lay-up techniques are generally processed by either a vacuum bag or autoclave technique. These processes are generally suitable for use with prepreg materials. Once again, the primary processing steps required are the top lay-ups, peel and bleeder plies preparation, and the bagging operation. The bagging film, which is trimmed to the size of the tool plate or platten, is placed over the prepared sample consisting of ply lay-ups as well as peel and bleeder plies. A vacuum connecter is attached to the bag and sealed appropriately to the tool plate. The system is then proofed by drawing a partial vacuum to ensure that the bag surface remains smooth relative to the prepared specimen surface. The complete assembly is then transferred to an oven for final curing under full vacuum following a prescribed thermal and cooling cycle. A schematic of the bag assembly processing is shown in Figure 1.18.

For the pressure bag process, an inflatable pressure bag is positioned within one or two glass fiber preforms; the assembly is then placed into a cold mold. After thorough impregnation with a resin, the pressure bag is inflated to approximately 35-45 psi and the part is cured against the mold, which is heated. After curing, the bag is deflated and removed from the part, typically through a mall opening at the end of the part.

Typical applications include seamless tanks for water softening, water handling and storage applications, filtration tanks and fire extinguisher tanks.

In the open molded vacuum bag process, a mold release film is placed over a fiber-glass resin lay-up. The joints are sealed with plastic and a vacuum is drawn. The resultant atmospheric pressure eliminates voids and forces out entrapped air and excess resin.

Applications include electronic components, radomes, and limited volume units.

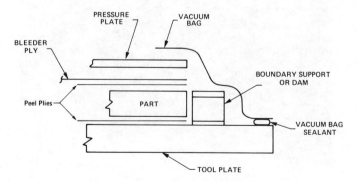

Figure 1.18. Vacuum Bagging/Autoclave Cure Method.

The counterpart to vacuum bagging is autoclave curing, which in essence provides a means for obtaining higher pressures and thus more dense materials. Since this technique requires the use of both pressure and temperature during the cure cycle, a vacuum system is also used. This ensure that excess air is removed during the processing while maintaining a controlled resin bleed-off. Generally, the pressure and temperature applied to the composite during the autoclave cure cycle are maintained over the entire cycle, and the applied vacuum may be applied over only some part of the cycle. Autoclave processing provides specimens of controlled thickness and minimum void content. Processing parameters associated with the technique consist of cure pressures ranging from 25 to 80 psi and cure temperatures up to 350 F. A photograph of a typical laboratory autoclave used in this processing technique is shown in Figure 1.19.

1.4.11. Matched Metal Die Molding

For producing small specimens of varying size and shape in large quantities it is often advantageous to prepare matched metal dies. These preforms can then be used under high pressures (up to 5000 psi) and extended temperatures (up to 1000°F) to form a variety of material component configurations. In addition, the forming environment can consist of a vacuum or inert gases, depending on the materials selected for processing. Such moldings are useful for forming both short and continuous fiber preforms. While this technique can be useful for producing standardized test specimens as used in data collection or quality control, a primary use is in the commercial sector where standard parts

Figure 1.19. Typical Autoclave.

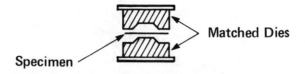

Figure 1.20. Matched Metal Die Molding Method.

of varying size and shape are (routinely) recurringly required. A schematic of the molding process for producing sheet elements is seen in Figure 1.20.

1.4.12. Rotational Molding

This method consists of using thermoplastic powders to which are added chopped fibers into a hollow heated mold. The mold is designed to rotate in one or two planes so that fusion of the mixed material will occur. After fusion the material is chilled and the resultant composite removed. This process has been used for producing containers and other shell type structural components.

1.4.13. Pultrusion

This technique relies upon fiber or fiber strands/rings which are impregnated with a resin binder and drawn through cast dies which shape the resultant stock. Cure is effecting by use of an oven using appropriate thermal cycling. Various types of structural shapes such as tubes and I beams are made in this manner.

1.4.14. Injection Molding

This process is used for producing composites with thermoplastic and thermosetting resins. For composites with thermoplastic matrices two feed systems are generally used with the matrix material introduced in pellet form and softened in an injection chamber. The heated material

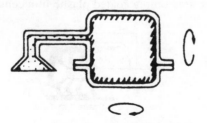

Figure 1.21. Rotational Molding Method.

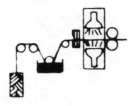

Figure 1.22. Continuous Pultrusion Method.

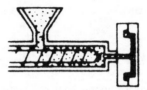

Figure 1.23. Injection Molding Method.

consisting of matrix and chopped fibers is then introduced into a cold mold chamber where the compounds used under heat pad pressure are generally on the order of one minute. Pressures may vary between 100 and 1500 psi, while cure temperatures range between 275 and 325°F. This process is useful for producing bulk molding compounds (BMC) as used in valves, business machine bearings, and small gears.

1.4.15. Centrifugal Casting

This process is useful for producing shell type pipe and container type structural elements. These round objects are formed by a centrifugal casting process in which a heated hollow pipe assembly containing chopped fibers is rotated and a resin mix introduced. The action of the centrifugal forces is such as to force the composite slowly against the pipe walls during the cure cycle. This cure cycle is accelerated by passing hot air through the inside of the pipe mandrel.

1.4.16. Sheet Molding Compounds

In this process as shown in Fig. 1.25 continuous strand rovings are chopped and deposited onto a coated plastic film consisting of a layer of

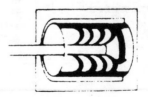

Figure 1.24. Centrifugal Casting Method.

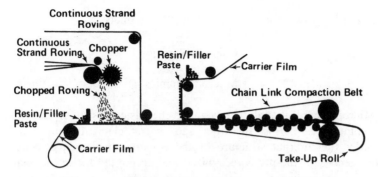

Figure 1.25. Chopped Fiber Processing: SMC.

resin filler paste, generally polyester. The paste and filler are rolled producing a sheet type product. The resultant material SMC (Sheet Molding Compound) is then cut and prepared for molding. Some applications of this technique are in the transport and business machine area.

1.5. Composite Material Behavior

In the previous sections we have defined and classified composite type materials as well as describing pertinent manufacturing processes associated with them. In this section we turn our attention to defining composites in terms of their response under applied loads. As a general approach to characterizing material behavior we can describe materials at the macroscopic level as being of the following * type:

Homogeneous/Isotropic	(HI)
Homogeneous/Anisotropic	(HA)
Nonhomogeneous/Isotropic	(NHI)
Nonhomogeneous/Anisotropic	(NHA)

A homogeneous body is one in which uniform properties exist throughout the body and for which the properties are not functionally dependent upon position dependence within the body. An isotropic body is one which has the same material properties in any direction at any point within the material medium.

Thus, due to the inherent heterogeneities and associated complexities of describing composite material behavior these materials are generally

---

* It should be noted that micromechanical analysis serves as a bridge between the two previously defined structural levels of composite classification, the macro- and micro-structural levels.

analyzed from two points of view Fig. 1.26: namely,

Micromechanics

Macromechanics

Micromechanical analysis recognizes the basic constituent elements of the composite, that is, the fiber and matrix elements, but does not consider the internal structure of the constituent elements. Thus, the heterogeneity of the ply is recognized and accounted for in the analysis. In order to develop a methodology suitable for characterizing material response at this level of analysis it is generally necessary to introduce a number of simplifying approximations. One of the most important of these concerns the fiber packing arrangement within the lamina so that the constituent properties along with the volume fraction of reinforcements can be used to predict the average properties of the lamina. These calculated values then serve as a basis for input into the analysis for predicting macrostructural properties.

Macromechanical analysis considers only the averaged properties of the lamina as being singularly important, that is, the microstructure of the lamina is ignored and the properties along and perpendicular to the fiber direction are recognized. The resultant structural element is considered to be built upon a systematic combination of these laminae with the structural element, be it beam, plate, or shell analyzed by classically formulated orthotropic theories.

This last point is a powerful concept in that the ability to structure and orient the lamina in prescribed sequences leads to several particu-

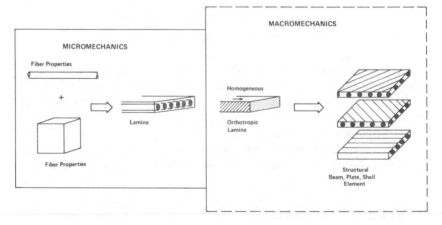

Figure 1.26. Analytical Characterization of Composite Materials.

larly significant advantages of composite materials when compared with conventional monolithic materials. Principal among these is the ability to tailor or match the lamina properties and orientation to the given structural loads. If a fibrous composite consists of a large number of laminae stacked sequentially, it is considered to be closely related to an isotropic material. Indeed, a simple series of plies stacked at 60° orientation with respect to one another leads to the synthesis of the material as isotropic in the plane of the ply arrangement.

The primary thrust of this book is focused upon the analytical methodology associated with characterizing composite material behavior at the macroscopic level. Associated with this level of material synthesis are certain response characteristics that distinguish composite materials from isotropic materials. Two important features characteristic of orthotropic materials are described in detail in the following chapter and relate to
– Normal and Shear Stress Coupling of Angle Ply Composites
– The Importance of Shear Stress Direction on Strength of Composites

1.6. Applications

The use of composite materials, that is, continuous or discontinuous fibers embedded in suitable metallic or nonmetallic matrices, is receiving even wider attention and use in commercial applications. Due to the inherent tailoring of properties of these materials, a number of unique design features can be utilized including such potential aspects as reduced weight, performance, increased service life and reduced system maintenance. As with the introduction of any new technology, certain problems associated with mass utilization arise. For composites such factors as introducing mechanized fabrication processes, providing design guidance for system fabrication, and establishing data bases for design are most important. Inherent in this utilization factor and data assembly is the role of the engineer himself. That is, there are educational implications of composite materials technology which require re-education of practicing engineers and educating new practitioners in terms of this new technology. Fundamentally, while engineers heretofore have been materials selectors it is now necessary for engineers to become materials designers. While economics will eventually dictate the overall usage of such materials, wider acceptance will be attained through increased user knowledge of composites as well as through the development of new fabrication processes.

These design features are realized in practice when the material constituents or structural elements are combined so as to achieve a desired performance superior to that of the individual constituents or

Table 1.1. Characteristics of Some Composite Materials.

Property	S-Glass	Boron	Graphite		Organic (Kevlar 49)
			High modulus	High strength	
Fabrication-handling					
Minimum bend radius	small	large	small	small	small
Susceptibility to handling damage	medium	medium	high	high	low
Machining characteristics	fair	poor	excellent	excellent	excellent
Fabric properties	fair	cannot be woven	good	good	excellent
Drape, as tape or fabric	good	poor	good	good	good
Other					
Cost	very low	high	moderate	moderate	moderate

[a] In plastic matrix.

structural elements. Thus, wider acceptance of composites will be attained through increased user knowledge of these materials as well as through the development of new cost-effective fabrication processes. Ultimately material availability and economics will dictate the role composites will play in engineering design. To this end, and as an example, in mass production there may be as much as a hundredfold increase in cost associated with composite materials used in mass transport vehicles such as automobiles as opposed to high-performance aircraft such as military aircraft. * Currently however, industry is embarked in implementation and usage of composites with recent developments including the advent of the all composite automobile ** and the equivalent in general aviation aircraft ***. The most widespread use of composite continues to be for aircraft structural components for both the airframe and engine components. Indeed, for the airframe it is anticipated that a 35% weight savings in existing metal structures can be achieved through use of composites and upwards of 70% in rudder designs. Since approximately one-third of an aircraft is structure, this weight savings can result in impressive performance gains. For engine components the converted weight savings from use of composites is equally impressive with potential for an order of magnitude of gain possible. Further fascinating developments using components as replacement and new design hardware for a variety of applications is forthcoming. The particular composites used will be based upon the characteristics of available composites as shown in the accompanying table, and the specific properties of high-performance filaments as described in the tables below.

---

* J.J. Broutman and R.H. Krock, Composite Materials, Vol. 3, Academic Press (1974).
** Newspaper Article
*** Astronautics and Aeronautics, May, 1983

Table 1.2. Characteristics of available composites

Type of Composite	High Strength/Weight Ratio	High Stiffness/Weight Ratio	Maximum Temperature Usage				Cryogenic Usage-Structural	Thermal Insulation	Chemical Resistance	Low Density
			−350°F	To −500°F	From 500–1000°F	Over 1000°F				
Fiber										
Glass-epoxy	X		X						X	
Glass-polyimide	X			X					X	
Kevlar-49-epoxy	X	X						X		X
Kevlar-49-polyimide	X	X						X		X
Graphite-epoxy	X	X	X				X			X
Graphite-polyimide	X	X					X			X
Boron-epoxy	X	X	X							
Boron-polyimide	X	X								
Boron-aluminum	X	X		X						
Tungsten-metal matrix						X				
Laminar										
Polymer coated fabrics			X					X	X	
Fiber-reinforced film			X					X	X	
Fiber-reinforced film			X							X
Honeycomb sandwich			X					X		X
Bi-metallics and inorganic-metalics		X			X	X	X	X		
Metalized organic film									X	
Skeletal										
Filled organic foam								X		
Filled metal matrix					X	X				
Filled honeycomb	X									
Particulate										
Dispersion-strengthed alloys					X	X				
Short fiber-strengthed alloys					X	X				
Fiber strengthed (ceramics)					X	X			X	
Cermets						X		X		X
Flake										
Mica-epoxy									X	

Table 1.3 Properties of High-Performance Filaments.

Property	S-Glass	Boron	Graphite		Organic (Kevlar 49)
			High modulus	High strength	
Mechanical					
Specific strength	high	high	moderate	high	very high
Specific modulus	low	high	very high	high	moderate
Uniformity of properties	excellent	excellent	moderate	moderate	excellent
Impact resistance	excellent	fair	poor	poor	excellent
Elongation to failure	high	low	low	moderate	moderate
Compression load-carrying ability [a]	low	very high	moderate	moderate	low
Erosion resistance [a]	moderate	high	low	low	high
Notched fatigue life [a]	low	moderate	moderate	moderate	high
Creep behavior [a]	high	very low	very low	low	moderate
Stress rupture strength	low	high	moderate	moderate	high
Shear bond to matrix	good	excellent	fair-good	good	fair-good
Thermal physical					
Density	high	high	moderate	moderate	low
Thermal coeff of expansion	high	moderate	very low	very low	very low
Thermal conductivity	low	moderate	high	high	low
Thermal stability	high	high	high	moderate	limited
Damping ability	good	fair	fair	fair-good	excellent
Dielectric properties	good	poor	poor	poor	excellent
Chemical					
Oxidation resistance	excellent	excellent	poor	poor	not established
Resistance to solvents	high	high	high	high	high
Resistance to high humidity [a]	fair	fair	poor	poor	excellent

1.7. Bibliography

• Books
1. R.H. Sonneborn, *Fiberglas Reinforced Plastics,* Reinhold, 1954.
2. R.F.S. Hearmon, *An Introduction to Applied Anisotropic Elasticity,* Oxford, 1961.
3. A.G. Leknitskii, *Theory of Elasticity for an Anisotropic Body,* Holden Day, 1963.
4. S.S. Oleesky, J.G. Mohr, *Handbook of Reinforced Plastics,* Reinhold, 1964.
5. L. Holliday, *Composite Materials,* Elsevier, 1966.
6. G.S. Hollister, C. Thomas, *Fibre Reinforced Materials,* Elsevier, 1966.
7. L.J. Broutman, R.H. Krock, *Modern Composite Materials,* Addison-Wesley, 1967.
8. R.M. Jones, *Mechanics of Composite Materials,* McGraw-Hill, 1967.
9. S.G. Leknitskii, *Anisotropic Plates,* Gordon Breach, 1968.
10. J. Burke, *Surfaces and Interfaces II,* Syracuse, 1968.
11. S.W. Tsai, J.C. Halpin, N.J. Pagano, *Composite Materials Workshop,* Technomic, 1968.
12. L.R. Calcote, *The Analysis of Laminated Composite Structures,* Van Nostrand, 1969.
13. J.E. Ashton, J.C. Halpin, P.H. Petit, *Primer on Composite Materials: Analysis,* Technomic, 1969.
14. G. Lubin, *Handbook of Fiberglass and Advanced Plastics Composites,* Van Nostrand, 1969.
15. J.E. Ashton, J.M. Whitney, *Theory of Laminated Plates,* Technomic, 1970.
16. A. Levitt, *Whisker Technology,* Wiley, 1970.
17. S.A. Ambartsumyan, *Theory of Anisotropic Plates,* Technomic, 1970.
18. R.L. McCullough, *Concepts of Fiber Resin Composites,* Marcel Dekker, 1971.
19. S.K. Garg, V. Svalbonas, G.A. Gurtman, *Analysis of Structural Composite Materials,* Marcel Dekker, 1973.
20. A. Kelly, *Strong Solids,* Clarendon, 1973.
21. J.A. Catherall, *Fibre Reinforcement,* Mills, Boon, 1973.
22. E. Scala, *Composite Materials for Combined Functions,* Hayden, 1973.
23. L.E. Nielson, *Mechanical Properties of Polymers and Composites,* Marcel Dekker, 1974.
24. J.R. Vinson, T.W. Chou, *Composite Materials and Their Use in Structures,* Applied Science, 1975.
25. L.J. Broutman, R.H. Krock, *Composite Materials, Vols. 1-8,* Academic Press, 1975.
26. V.K. Tewary, *Mechanics of Fibre Composites,* Halsted, 1978.
27. R.M. Christensen, *Mechanics of Composite Materials,* Wiley, 1979.
28. B.D. Agarwal, L.J. Broutman, *Analysis and Performance of Fiber Composites,* John Wiley, 1980.
29. S.W. Tsai, H.T. Hahn, *Introduction to Composite Materials,* Technomic, 1980.
30. E. Lenoe, D. Oplinger, J. Burke, *Fibrous Composites in Structural Design,* Plenum, 1980.
31. J.M. Whitney, I.M. Daniel, R.B. Pipes, *Experimental Mechanics of fiber Reinforced Composite Materials,* Technomic, 1982.
32. M.M. Schwartz, *Composite Materials Handbook,* McGraw-Hill, 1984.
• Publications
33. ASTM STP 427, *Fiber Strengthened Metallic Composites,* 1967.
34. ASTM STP 438, *Metal Matrix Composites,* 1967.
35. ASTM STP 452, *Interfaces in Composites,* 1968.
36. ASTM STP 460, 497, 546, 617 and 674, *Composite Materials: Testing and Design,* 1969, 1971, 1973, 1977 and 1979 respectively.
37. ASTM STP 521, *Analysis of Test Methods for High Modulus Fibers and Composites,* 1973.
38. ASTM STP 524, *Application of Composite Materials,* 1973.
39. ASTM STP 568, *Foreign Object Impact Damage to Composites,* 1975.
40. ASTM STP 569, *Fatigue of Composites,* 1975.

41. ASTM STP 580, *Reliability*, 1975.
42. ASTM STP 593, *Fracture Mechanics*, 1976.
43. ASTM STP 602, *Environmental Effects*, 1976.
44. ASTM STP 638, *Fatigue*, 1977.
45. *Advanced Composite Design Guide*, Five Volumes, Air Force Materials Laboratory, Third Edition, 1977.
46. ASTM STP 636, *Fatigue of Filamentary Composite Materials*
47. ASTM STP 658, *Advanced Composite Materials - Environmental Effects*, 1978.
48. ASTM STP 696, *Nondestructive Evaluation and Flaw Criticality for Composite Materials*, 1979.
49. ASTM STP 704, *Commercial Opportunities for Advanced Composites*, 1980.
50. ASTM STP 734, *Test Methods and Design Allowables for Fibrous Composites*, 1981.
51. ASTM STP 749, *Joining of Composite Materials*, 1981.
52. ASTM STP 768, *Composites for Extreme Environments*, 1982.
53. ASTM STP 772, *Short Fiber Reinforced Composite Materials*, 1982.
54. ASTM STP 775, *Damage in Composite Materials*, 1982.
55. ASTM STP 787, *Composite Materials: Testing and Design (6th Conference)*, 1982.
56. ASTM STP 797, *Composite Materials: Quality Assurance and Processing*.
57. ASTM STP 831, *Long-Term Behavior of Composites*, 1983.
• Journals
 1. J. of Composite Materials
 2. J. of Engr. Matls. and Technology (ASME)
 3. J. Applied Mechanics (ASME)
 4. J. Testing and Evaluation (ASTM)
 5. J. Mechanics and Physics of Solids
 6. J. of Materials Science (British)
 7. AIAA Journal
 8. International Journal of Solids and Structures
 9. Experimental Mechanics (SESA)
10. Shock and Vibration Digest
11. Composites (British)
12. Polymer Engineering and Science
13. Fibre Science and Technology (British)
14. Composites Technology Review (ASTM)
15. Polymer Composites (New Journal by SPE)
16. Mechanics of Materials (Ed. S. Nemat-Nasser, North Holland Publ. Co.)
17. International Journal of Engineering Science
18. International Journal of Composite Structures (British)
19. Journal of Sound and Vibration.

2. Anisotropic Elasticity and Laminate Theory

2.1. Introduction

In Chapter 1, composite materials were discussed as well as some of the means for fabricating structural components composed of these composites. Today, the preponderance of uses for composite materials is in the form of laminated beam, plate and shell structural members. These will be discussed in some detail in Chapters 3, 4 and 5.

However, before understanding the physical behavior of composite plates and shells, and before being able to quantitatively determine the stresses, strains, deformations, natural frequencies, and buckling loads in them, a clear understanding of anisotropic elasticity is mandatory.

Most engineers and material scientists are well versed in the behavior of and use of isotropic materials, such as most metals and pure polymers. An isotropic material has identical mechanical, physical, thermal and electrical properties in every direction. Materials exhibiting directional characteristics are called anisotropic.

In general, isotropic materials are mathematical approximations to the true situation. For instance, in polycrystalline metals, the structure is usually made up of numerous anisotropic grains, wherein macroscopic isotropy exists in a statistical sense only when individual grains are randomly oriented. However, the same materials could be macroscopically anisotropic due to cold working, forging or spinning during a fabrication process. Other materials such as wood, human and animal bone, and all fiber reinforced materials such as those of Chapter 1 are naturally anisotropic.

As discussed in Chapter 1, fiber-reinforced composite materials are uniquely useful because the use of long fibers results in a material which has a higher strength-to-density ratio and/or stiffness-to-density ratio than any other material system at moderate temperature, and there exists the opportunity to uniquely tailor the fiber orientations to a given geometry, applied load and environmental system. For short fiber composites, used mainly in high production, low cost systems, the use of fibers makes the composites competitive and superior to the plastic and metal alternatives. Finally, the use of two or more fibers with one matrix is an exciting new dimension to be exploited where one fiber is stronger or stiffer while the other fiber is less expensive but desirable for less sensitive locations in an overall structural component. So through the use of composite materials, the engineer is not merely a materials selector, but he is also a materials designer.

There are many texts and reference books dealing with thin walled structures (beams, plates and shells) such as Shames [1], Vinson [2], Timoshenko and Woinowsky-Krieger [3], and Marguerre and Woernle [4]. From those it is seen that for the small deflection, linear elastic analysis of such structures, the equilibrium equations, strain-displacement relations, and compatibility equations remain the same whether the structure is composed of an isotropic material or an anisotropic composite material. However, it is very necessary to drastically alter the stress-strain relations, also called the constitutive relations, to account for the anisotropy of the composite materials system.

A quantitative understanding of the value of using composite materials in a structure is seen through deriving systematically the anisotropic elasticity tensor matrix, as will be done in Section 2.2 below.

2.2. Derivation of the Anisotropic Elastic Stiffness and Compliance Matrices

Consider an elastic solid body of any general shape, and assume it is composed of an infinity of *material points* within it. In order to deal with a *continuum*, we also assume that the material points are infinitely large compared to the molecular lattice spacing of the particular material. If one assigns a cartesian reference frame to the elastic body shown in Figure 2.1, one then calls this rectangular parallelpiped material point a *control element* of dimension dx, dy and dz.

On the surface of the control element there can exist both normal stresses (those perpendicular to the plane of the face) and shear stresses (those parallel to the plane of the face). On any one face the three mutually orthogonal stress components comprise a vector, which is called a *surface traction*.

It is important to note the sign convention and the meaning of the subscripts of these surface stresses. For a stress component on a face whose outward normal is in the direction of a positive axis, the stress component is positive when it is in the direction of a positive axis. Also, when a stress component is on a face whose outward normal is in the direction of a negative axis, the stress component is positive when it is in the direction of a negative axis. This can be seen clearly in Figure 2.1.

The first subscript of any stress component on any face of the control element signifies the axis to which the outward normal of the face is

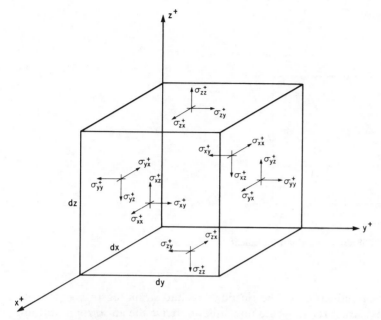

Figure 2.1. Positive Stresses on a Control Element of an Elastic Body.

parallel; the second subscript refers to the axis to which that stress component is parallel. Again, see Figure 2.1 above.

The strains occurring in an elastic body have the same subscripts as the stress components but are of two types. Dilitational or extensional strains are denoted by ϵ_{ii}, where $i = x$, y, z, and are a measure of the change in dimension of the control volume in the subscripted direction due to normal stresses, σ_{ii}, acting on the control volume. Shear strains, ϵ_{ij} $(i \neq j)$ are proportional to the change in angles of the control volume from 90°, changing the rectangular control volume into a parallelpiped due to the shear stresses, σ_{ij}, $i \neq j$. For example, looking at the control volume x-y plane shown in Figure 2.2 below, shear stresses σ_{xy} and σ_{yx} cause the square control element with 90° corner angles to become a parallelogram with the corner angle ϕ as shown. Here, the change in angle γ_{xy} is

$$\gamma_{xy} = \frac{\pi}{2} - \phi.$$

The shear strain ϵ_{xy}, a tensor quantity is defined by

$$\epsilon_{xy} = \gamma_{xy}/2.$$

Similarly, $\epsilon_{xz} = \gamma_{xz}/2$, and $\epsilon_{yz} = \gamma_{yz}/2$.

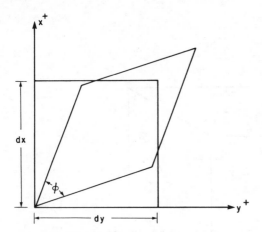

Figure 2.2. Shearing of a Control Element.

Having defined all of the elastic stress and strain tensor components, the stress-strain relations are now used to derive the anisotropic stiffness and compliance matrices.

The following derivation of the stress strain relations for an anisotropic material parallel the derivation of Sokolnikoff [5], and Vinson and Chou [6]. Although the derivation is very formal mathematically to the reader who is primarily interested with the end result, the systematic derivation provides confidence in the extended use of the results.

From knowledge of basic strength of materials [1], both stress, σ_{ij}, and strain ϵ_{ij}, are second order tensor quantities, where in three dimensional space they have $3^2 = 9$ components. They are equated by means of a fourth order tensor quantity, C_{ijkl}, which therefore has $3^4 = 81$ components, with the resulting constitutive equation:

$$\sigma_{ij} = C_{ijkl}\epsilon_{kl} \tag{2.1}$$

where i, j, k and l assume values of 1, 2, 3 or x, y, z in the commonly defined Cartesian coordinate system. Fortunately, there is no actual material which has eighty-one elastic constants. Both the stress and strain tensors are symmetric, i.e., $\sigma_{ij} = \sigma_{ji}$ and $\epsilon_{kl} = \epsilon_{lk}$, and therefore the following shorthand notation may be used:

$$
\begin{array}{llll}
\sigma_{11} = \sigma_1 & \sigma_{23} = \sigma_4 & \epsilon_{11} = \epsilon_1 & 2\epsilon_{23} = \epsilon_4 \\
\sigma_{22} = \sigma_2 & \sigma_{31} = \sigma_5 & \epsilon_{22} = \epsilon_2 & 2\epsilon_{31} = \epsilon_5 \\
\sigma_{33} = \sigma_3 & \sigma_{12} = \sigma_6 & \epsilon_{33} = \epsilon_3 & 2\epsilon_{12} = \epsilon_6
\end{array}
\tag{2.2}
$$

Using (2.2), (2.1) can be written as follows:

$$\sigma_1 = C_{11}\epsilon_1 + C_{12}\epsilon_2 + C_{13}\epsilon_3 + C_{14}\epsilon_4 + C_{15}\epsilon_5 + C_{16}\epsilon_6$$

$$\vdots \qquad\qquad\qquad\qquad\qquad\qquad \vdots \qquad (2.3)$$

$$\sigma_6 = C_{61}\epsilon_1 + C_{62}\epsilon_2 + C_{63}\epsilon_3 + C_{64}\epsilon_4 + C_{65}\epsilon_5 + C_{66}\epsilon_6$$

It should be noted that the contracted C_{ij} quantities in (2.3) are not tensor quantities, and therefore cannot be transformed as such.

Hence by the symmetry in the stress and strain tensors the elasticity tensor immediately reduces to the 36 components displayed in (2.3). In addition, if a strain energy function, W, exists (see Reference 2–6), i.e.,

$$W = \tfrac{1}{2}\sigma_{ij}\epsilon_{ij}$$

in such a way that

$$\frac{\partial W}{\partial \epsilon_{ij}} = C_{ijkl}\epsilon_{kl} = \sigma_{ij} \qquad (2.4)$$

then the independent components of C_{ijkl} are reduced to 21, since $C_{ijkl} = C_{klij}$ or now it can be written $C_{ij} = C_{ji}$.

Next, to simplify the general mathematical anisotropy to the cases of very practical importance, consider the Cartesian coordinate system only. (However, the results are applicable to any curvilinear orthogonal coordinate system of which there are twelve, some of which are spherical, cylindrical, elliptical.).

First, consider an elastic body whose properties are symmetric with respect to the $X_1 - X_2$ plane. The resulting symmetry can be expressed by the fact that the C_{ij}'s discussed above must be invariant under the transformation $X_1 = X_1'$, $X_2 = X_2'$ and $X_3 = -X_3'$, shown in Figure 2.3 below.

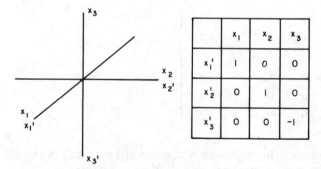

Figure 2.3. One Plane of Symmetry.

Also shown above in the table are the direction cosines, t_{ij}, associated with this transformation. The stresses and strains of the primed coordinate system are related to those of the original (unprimed) coordinate system by the well-known relationships:

$$\sigma'_{\alpha\beta} = t_{\alpha i} t_{\beta j} \sigma_{ij} \quad \text{and} \quad \epsilon'_{\alpha\beta} = t_{\alpha i} t_{\beta j} \epsilon_{ij}.$$

Therefore, for $i = 1, 2, 3, 6$, $\sigma'_i = \sigma_i$ and $\epsilon'_i = \epsilon_i$, i.e., $\sigma'_{11} = t_{11} t_{11} \sigma_{11} = \sigma_{11}$. However, from the direction cosines, $\epsilon'_{23} = -\epsilon_{23}$ or $\epsilon'_4 = -\epsilon_4$, and $\sigma'_4 = -\sigma_4$; likewise $\epsilon'_{31} = -\epsilon_{31}$, hence $\epsilon'_5 = -\epsilon_5$ and $\sigma'_5 = -\sigma_5$. For example, $\sigma'_{23} = \sigma'_4 = t_{22} t_{33} \sigma_{23} = (1)(-1) \sigma_{23} = -\sigma_{23} = -\sigma_4$.

If one looks in detail at (2.3) then, $\sigma'_4 = C_{41} \epsilon'_1 + C_{42} \epsilon'_2 + C_{43} \epsilon'_3 + C_{44} \epsilon'_4 + C_{45} \epsilon'_5 + C_{46} \epsilon'_6$, $\sigma_4 = C_{41} \epsilon_1 + C_{42} \epsilon_2 + C_{43} \epsilon_3 + C_{44} \epsilon_4 + C_{45} \epsilon_5 + C_{46} \epsilon_6$. It is clearly seen from these two equations that $C_{41} = C_{42} = C_{43} = C_{46} = 0$. From other similar examinations, it is seen also that $C_{25} = C_{35} = C_{64} = C_{65} = 0$, $C_{51} = C_{52} = C_{53} = C_{56} = 0$, and $C_{14} = C_{15} = C_{16} = C_{24} = C_{34} = 0$.

So, for a material having only *one* plane of symmetry the number of elastic constants is reduced to 13. Note that from a realistic engineering point of view this would require thirteen independent physical tests (at each temperature and humidity condition!) – an almost impossible task both manpower wise and budget wise.

Now, materials (real materials) which have three mutually orthogonal planes of elastic symmetry are called "orthotropic" (a shortened term for orthogonally anisotropic). In that case, other terms in the elasticity matrix are also zero, namely

$$C_{16} = C_{26} = C_{36} = C_{45} = 0.$$

Therefore, the elasticity tensor for orthotropic materials is shown below, remembering that $C_{ij} = C_{ji}$,

$$C_{ij} = \begin{bmatrix} C_{11} & C_{12} & C_{13} & 0 & 0 & 0 \\ C_{21} & C_{22} & C_{23} & 0 & 0 & 0 \\ C_{31} & C_{32} & C_{33} & 0 & 0 & 0 \\ 0 & 0 & 0 & C_{44} & 0 & 0 \\ 0 & 0 & 0 & 0 & C_{55} & 0 \\ 0 & 0 & 0 & 0 & 0 & C_{66} \end{bmatrix}. \tag{2.5}$$

So, for orthotropic elastic bodies, such as composite materials in three dimensional configurations, there are nine elasticity constants.

Hence, with (2.5) and (2.1), the explicit stress strain relations for an orthotropic, three dimensional material are: $\sigma_i = C_{ij}\epsilon_j$ $(i, j = 1, 2, \ldots, 6)$ or more explicity,

$$
\begin{aligned}
\sigma_1 &= C_{11}\epsilon_1 + C_{12}\epsilon_2 + C_{13}\epsilon_3 \\
\sigma_2 &= C_{21}\epsilon_1 + C_{22}\epsilon_2 + C_{23}\epsilon_3 \\
\sigma_3 &= C_{31}\epsilon_1 + C_{32}\epsilon_2 + C_{33}\epsilon_3 \\
\sigma_4 &= \sigma_{23} = C_{44}\epsilon_4 = 2C_{44}\epsilon_{23} \\
\sigma_5 &= \sigma_{31} = C_{55}\epsilon_5 = 2C_{55}\epsilon_{31} \\
\sigma_6 &= \sigma_{12} = C_{66}\epsilon_6 = 2C_{66}\epsilon_{12}
\end{aligned}
\tag{2.6}
$$

If (2.6) is inverted, then, through standard matrix transformation:

$$
\begin{aligned}
\epsilon_1 &= a_{11}\sigma_1 + a_{12}\sigma_2 + a_{13}\sigma_3 \\
\epsilon_2 &= a_{21}\sigma_1 + a_{22}\sigma_2 + a_{23}\sigma_3 \\
\epsilon_3 &= a_{31}\sigma_1 + a_{32}\sigma_2 + a_{33}\sigma_3 \\
\epsilon_4 &= 2\epsilon_{23} = a_{44}\sigma_{23} = a_{44}\sigma_4 \\
\epsilon_5 &= 2\epsilon_{31} = a_{55}\sigma_{31} = a_{55}\sigma_5 \\
\epsilon_6 &= 2\epsilon_{12} = a_{66}\sigma_{12} = a_{66}\sigma_6.
\end{aligned}
\tag{2.7}
$$

The a_{ij} matrix, is called the compliance matrix, is the transpose of the cofactor matrix of the C_{ij}'s divided by the determinant of the C_{ij} matrix and each term is defined as

$$
a_{ij} = \frac{\left[C_0 C_{ij} \right]^T}{|C_{ij}|}.
\tag{2.8}
$$

Again, the a_{ij} quantities are not tensors, and should not be transformed as such. In fact, factors of 1, 2 and 4 appear when relating the tensor compliance quantities a_{ijkl} and the contracted compliance quantities a_{ij}.

It can be easily shown that $a_{ij} = a_{ji}$ and

$$
\epsilon_i = a_{ij}\sigma_j \quad (\text{where } i, j = 1, 2, \ldots, 6).
\tag{2.9}
$$

Table 2.1 is useful to see the number of elastic coefficients present in both two and three dimensional elastic bodies.

Table 2.1
Summary of the number of elastic coefficients involved for certain classes of materials

Class of Material	Number of nonzero coefficients	Number of independent coefficients
Three-Dimensional Case		
General anisotropic	36	21
One-plane of symmetry	20	13
Two-planes of symmetry	12	9
Transversely isotropic	12	5
Isotropic	12	2
Two-Dimensional Case		
General anisotropic	9	6
One-plane of symmetry	9	6
Two-planes of symmetry	5	4
Transversely isotropic	5	4
Isotropic	5	2

2.3. The Physical Meaning of the Components of the Orthotropic Elastic Tensor.

So far, the components of both the stiffness matrix, C_{ij}, and the compliance matrix, a_{ij}, are mathematical symbols to relate stresses and strains. By performing simple tensile and shear tests all of the components above can be related to physical or mechanical properties.

Consider a simple, standard tensile test in the X_1 direction. The resulting stress and strain tensors are

$$\sigma_{ij} = \begin{bmatrix} \sigma_{11} & 0 & 0 \\ 0 & 0 & 0 \\ 0 & 0 & 0 \end{bmatrix}, \quad \epsilon_{ij} = \begin{bmatrix} \epsilon_{11} & 0 & 0 \\ 0 & -\nu_{12}\epsilon_{11} & 0 \\ 0 & 0 & -\nu_{13}\epsilon_{11} \end{bmatrix} \quad (2.10)$$

where the Poisson's ratio, ν_{ij}, is very carefully defined as the negative of the ratio of the strain in the X_j direction to the strain in the X_i direction due to an applied stress in the X_i direction. In other words in the above it is seen that $\epsilon_{22} = -\nu_{12}\epsilon_{11}$ or $\nu_{12} = -\epsilon_{22}/\epsilon_{11}$.

Also, the constant of proportionality between stress and strain is noted to be E_i, the modulus of elasticity in the X_i direction. Thus,

$$\epsilon_1 = a_{11}\sigma_1 = \frac{\sigma_1}{E_1}$$

$$\epsilon_2 = a_{21}\sigma_1 = -\nu_{12}\epsilon_1 = -\frac{\nu_{12}\sigma_1}{E_1}$$

$$\epsilon_3 = a_{31}\sigma_1 = -\nu_{13}\epsilon_1 = -\frac{\nu_{13}\sigma_1}{E_1}$$

Therefore,

$$a_{11} = 1/E_1, \quad a_{21} = -\nu_{12}/E_1, \quad a_{31} = -\nu_{13}/E_1. \tag{2.11}$$

For a simple tensile test in the X_2 direction, it is found that

$$a_{12} = -\nu_{21}/E_2, \quad a_{22} = 1/E_2, \quad a_{32} = -\nu_{23}/E_2. \tag{2.12}$$

Likewise, a tensile test in the X_3 direction yields

$$a_{13} = -\nu_{31}/E_3, \quad a_{23} = -\nu_{32}/E_3, \quad a_{33} = 1/E_3. \tag{2.13}$$

From the fact that $a_{ij} = a_{ji}$, it is seen that

$$\frac{\nu_{ij}}{E_i} = \frac{\nu_{ji}}{E_j} \quad (i, j = 1, 2, 3). \tag{2.14}$$

Next, consider a hypothetical simple shear test as shown in Figure 2.4. In this case the stress, strain, and displacement tensor components

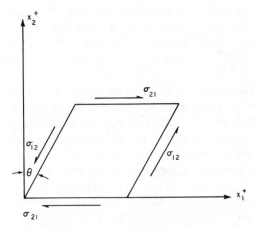

Figure 2.4. Shear Stresses and Strains.

are:

$$\sigma_{ij} = \begin{bmatrix} 0 & \sigma_{12} & 0 \\ \sigma_{21} & 0 & 0 \\ 0 & 0 & 0 \end{bmatrix}, \quad \epsilon_{ij} = \begin{bmatrix} 0 & \epsilon_{12} & 0 \\ \epsilon_{21} & 0 & 0 \\ 0 & 0 & 0 \end{bmatrix}, \quad U_{i,j} = \begin{bmatrix} 0 & 0 & 0 \\ \sigma_{21}/G_{21} & 0 & 0 \\ 0 & 0 & 0 \end{bmatrix}.$$

In the above, U_i is the displacement and $U_{i,j} \equiv (\partial U_i)/(\partial X_j)$. From elementary strength of materials the constant of proportionality between the shear stress σ_{21} and the angle θ is G_{21}, the shear modulus in the $X_1 - X_2$ plane.

From the theory of elasticity

$$\epsilon_{12} = \tfrac{1}{2}(U_{1,2} + U_{2,1}) = \frac{\sigma_{21}}{2G_{21}} = \frac{\tan \theta}{2}. \tag{2.15}$$

From (2.7), $\epsilon_6 = a_{66}\sigma_6$, or

$$\epsilon_{12} = \frac{a_{66}\sigma_{21}}{2} = \frac{\sigma_{21}}{2G_{21}}.$$

Hence,

$$a_{66} = \frac{1}{G_{21}} = \frac{1}{G_{12}}. \tag{2.16}$$

Similarly,

$$a_{44} = \frac{1}{G_{23}} \quad \text{and} \quad a_{55} = \frac{1}{G_{13}}. \tag{2.17}$$

Thus, all a_{ij} components have now been related to mechanical properties, and it is seen that to characterize a three dimensional orthotropic body, nine physical quantities – hence nine separate tests – are needed (that is E_1, E_2, E_3, G_{12}, G_{23}, G_{31}, ν_{12}, ν_{13}, ν_{21}, ν_{23}, ν_{31} and ν_{32}, and equation (2.14)).

There are several sets of equations for obtaining the composite elastic properties from those of the fiber and matrix materials. These include those of Halpin and Tsai [7], Hashin [8], and Christensen [9]. Recently, Hahn [10] codified certain results for fibers of circular cross section which are randomly distributed in a plane normal to the unidirectionally oriented fibers. For that case the composite is macroscopically transversely isotropic, that is $(\ )_{12} = (\ )_{13}$, $(\ )_{22} = (\ )_{33}$ and $(\ )_{44} = (\ )_{55}$, where the parentheses could be E, G or ν; hence, the elastic properties involve only five independent constants, namely $(\ )_{11}$, $(\ )_{22}$, $(\ )_{12}$, $(\ )_{44}$ and $(\ )_{66}$.

For several of the elastic constants, Hahn states that they all have the same functional form:

$$P = \frac{(P_f V_f + \eta P_m V_m)}{(V_f + \eta V_m)}$$

(2.18)

where for the elastic constant P, the P_f, P_m and η are given in the table below, and where V_f and V_m are the volume fractions of the fibers and matrix respectively (and whose sum equals unity):

Elastic Constant	P	P_f	P_m	η
E_{11}	E_{11}	E_{11f}	E_m	1
ν_{12}	ν_{12}	ν_{12f}	ν_m	1
G_{12}	$1/G_{12}$	$1/G_{12f}$	$1/G_m$	η_6
G_{23}	$1/G_{23}$	$1/G_{23f}$	$1/G_m$	η_4
K_T	$1/K_T$	$1/K_f$	$1/K_m$	η_K

The expressions for E_{11} and ν_{12} are called the Rule of Mixtures. In the above K_T is the plane strain bulk modulus, $K_f = [E_f/2(1 - \nu_f)]$ and $K_m = [E_m/2(1 - \nu_m)]$. Also, the η's are given as follows:

$$\eta_6 = \frac{1 + G_m/G_{12f}}{2}$$

$$\eta_4 = \frac{3 - 4\nu_m + G_m/G_{23f}}{4(1 - \nu_m)}$$

$$\eta_K = \frac{1 + G_m/K_f}{2(1 - \nu_m)}.$$

The shear modulus of the matrix material, G_m, if isotropic is given by $G_m = E_m/2(1 + \nu_m)$.

The transverse moduli of the composite, $E_{22} = E_{33}$, are found from the following equation:

$$E_{22} = E_{33} = \frac{4K_T G_{23}}{K_T + mG_{23}}$$

(2.19)

where

$$m = 1 + \frac{4K_T \nu_{12}^2}{E_{11}}.$$

The equations above can be used for composites reinforced with anisotropic fibers such as graphite and aramid (Kevlar) fibers. If the fibers are isotropic, simplification of the above equations will result. In that case also η_K becomes

$$\eta_K = \frac{1 + (1 - 2\nu_f) G_m/G_f}{2(1 - \nu_m)}.$$

Hahn notes that for most structural composites, $G_m/G_f < 0.05$. Thus the η parameters are approximately:

$$\eta_6 \approx 0.5; \quad \eta_4 = \frac{3 - 4\nu_m}{4(1 - \nu_m)}; \quad \eta_K = \frac{1}{2(1 - \nu_m)}.$$

Finally, noting that $\nu_m = 0.35$ for most epoxies, then $\eta_4 = 0.662$ and $\eta_K = 0.77$.

The above equations along with (2.14) provide the engineer with the wherewithal to estimate the elastic constants for a composite material, if the constituent properties and volume fractions are known.

2.4. Thermal and Hygrothermal Considerations

In the previous two sections, the elastic relations developed pertain only to an anisotropic elastic body at one temperature, that temperature being the "stress free" temperature, that is the temperature at which the body is considered to be free of stress if it is under no mechanical static or dynamic loadings.

However, in both metallic and composite structures changes in temperature are commonplace both during fabrication and during structural usage. Changes in temperature result in two effects that are very important. First, most materials expand when heated, and contract when cooled, and in most cases this expansion is proportional to the temperature change. If, for instance, one had a long thin bar of a given material then with change in temperature, the ratio of the change in length of the bar, ΔL, to the original length L, is related to the temperature of the bar T, as shown in Figure 2.5.

Mathematically, this can be written as

$$\epsilon_{\text{THERMAL}} = \frac{\Delta L}{L} = \alpha \Delta T \tag{2.20}$$

where α is the coefficient of thermal expansion, i.e., the proportionality constant between the "thermal" strain ($\Delta L/L$) and the change in temperature, ΔT, from some reference temperature at which there are no thermal stresses or thermal strains.

The second major effect of temperature change relates to stiffness and strength. Most materials become softer, more ductile, and weaker as they

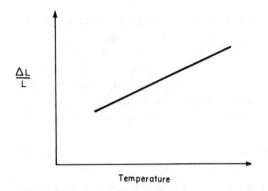

Figure 2.5. Change in Length of a Bar or Rod as a Function of Temperature.

are heated. Typical plots of ultimate strength, yield stress and modulus of elasticity with temperature are shown in Figure 2.6. In performing a stress analysis, determining the natural frequencies, or the buckling load of a heated or cooled structure one must use the strengths and the moduli of elasticity of the material at the temperature at which the structure is expected to perform.

In an orthotropic material, such as a composite, there will or could be three different coefficients of thermal expansion, and three different thermal strains, one in each of the orthogonal directions comprising the orthotropic material (equation 2.20 would then have subscripts of 1, 2 and 3 on both the strains and the coefficients of thermal expansion).

During the mid-seventies another physical phenomenon associated with polymer matrix composites was recognized as important. It was found that the combination of high temperature and high humidity caused a doubly deleterious effect on the structural performance of these composites. Engineers and material scientists became very concerned about these effects, and considerable research effort was expended in

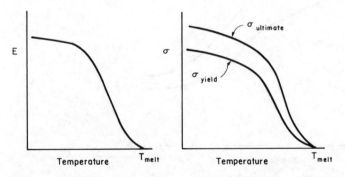

Figure 2.6. Modulus of Elasticity and Strengths as Functions of Temperature.

42

studying this new phenomenon. Conferences [11] were held which discussed the problem, and both short range and long range research plans were proposed. The twofold problem involves the fact that the combination of high temperature and high humidity results in the entrapment of moisture in the polymer matrix, with attendant weight increase ($\leqslant 2\%$) and more importantly, a swelling of the matrix. It was realized [12] that the ingestion of moisture varied linearly with the swelling so that in fact

$$\epsilon_{\text{HYGROTHERMAL}} = \frac{\Delta L}{L} = \beta \Delta m \tag{2.21}$$

where Δm is the increase from zero moisture measured in percentage weight increase, and β is the coefficient of hygrothermal expansion, analogous to the coefficient of thermal expansion, depicted in equation (2.20). This analogy is a very important one because one can see that the hygrothermal effects are entirely analogous mathematically to the thermal effect. The reader, however, should be aware that there exist some ambiguities in the literature regarding the coefficient of hygrothermal expansion.

The second effect also is similar to the thermal effect. These are shown qualitatively in Figure 2.7. Dry polymers have good properties until a temperature is reached, traditionally called by polymer chemists the "glass transition temperature," above which both strength and stiffness deteriorate rapidly. If the same polymer is saturated with moisture, not only are the mechanical properties degraded at any one temperature but the glass transition temperature for that polymer is significantly lower.

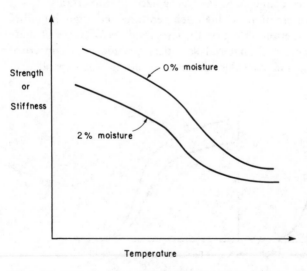

Figure 2.7. Mechanical Properties as a Function of Temperature and Moisture Absorption.

Thus, for modern polymer matrix composites one must include not only the thermal effects but also the hygrothermal effects or the structure could be considerably underdesigned.

Thus, to deal with the real world of polymer composites, equation (2.9) must be modified to read

$$\epsilon_i = a_{ij}\sigma_j + \alpha_i \Delta T + \beta_i \Delta m \quad (i = 1, 2, 3) \tag{2.22}$$

$$\epsilon_i = a_{ij}\sigma_j \quad (i = 4, 5, 6) \tag{2.23}$$

where in each equation $j = 1 - 6$. $\tag{2.24}$

Two types of equations are shown above because both thermal and hygrothermal effects are dilatational only, that is, they cause an expansion or contraction, but do not affect the shear stresses or strains. This is important to remember.

2.5. Laminae of Composite Materials

Almost all practical composite material structures are thin in the thickness direction because their superior properties permit the use of thin-walled structures. Most polymeric matrix composites are made in the form of a uniaxial set of fibers surrounded by a polymeric matrix in the form of a tape several inches wide termed as "prepreg." The basic element in most long fiber composite structures is a lamina of fiber plus matrix, all fibers oriented in one direction, made by laying the prepreg tape of a certain length side by side. In the next section, stacking of various laminae to form a superior structure termed a laminate will be discussed.

To effect this, consider a small element of a lamina of constant thickness h, wherein the principal material axes are labelled 1 and 2, that is 1 direction is parallel to the fibers, the 2 direction is normal to them, and consider that the beam, plate or shell geometric axes are x and y as depicted in Figure 2.8.

This element has the positive directions of all stresses shown in a consistent manner [2,5,6]. If one does a force equilibrium study to relate σ_x, σ_y and σ_{xy} to σ_1, σ_2, and σ_{12}, it is exactly analogous to the Mohr's circle analysis in basic strength of materials with the result that, in matrix form,

$$\begin{bmatrix} \sigma_1 \\ \sigma_2 \\ \sigma_6 \end{bmatrix} = [T]_{CL} \begin{bmatrix} \sigma_x \\ \sigma_y \\ \sigma_{xy} \end{bmatrix} \tag{2.25}$$

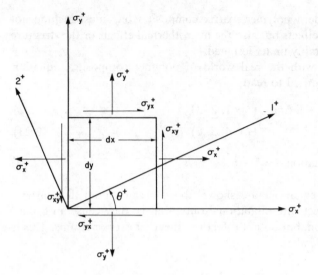

Figure 2.8. Laminae Coordinate Systems.

where

$$[T]_{CL} = \begin{bmatrix} m^2 & n^2 & +2mn \\ n^2 & m^2 & -2mn \\ -mn & mn & (m^2 - n^2) \end{bmatrix}$$ (2.26)

where $m = \cos\theta$, $n = \sin\theta$, and θ is defined positive as shown in Figure 2.8, and where the subscripts CL refer to the classical two-dimensional case only, that is, in the 1-2 plane or the x-y plane only.

Analogously, a strain relationship also follows for the classical isothermal case

$$\begin{bmatrix} \epsilon_1 \\ \epsilon_2 \\ \epsilon_{12} \end{bmatrix} = [T]_{CL} \begin{bmatrix} \epsilon_x \\ \epsilon_y \\ \epsilon_{xy} \end{bmatrix}.$$ (2.27)

However, these classical two-dimensional relationships must be modified to treat a composite material to include thermal effects, hygrothermal effects, and the effects of transverse shear deformation treated in detail elsewhere [6,13–15]. The effects of transverse shear deformation, shown through the inclusion of the $\sigma_4 - \epsilon_4$ and $\sigma_5 - \epsilon_5$ relations shown below in equations (2.28) and (2.30), must be included in composite materials, because in the fiber direction the composite has many of the mechanical properties of the fiber itself (strong and stiff) while in the thickness direction the fibers are basically ineffective and the shear properties are dominated by the weaker matrix material. Similarly, because quite often the matrix material has much higher coefficients of thermal and hygrothermal expansion (α and β), thickening of the thin lamina cannot be ignored in some cases. Hence, without undue derivation, the equations (2.25) through (2.27) are modified to be:

$$
\begin{bmatrix} \sigma_1 \\ \sigma_2 \\ \sigma_3 \\ \sigma_4 \\ \sigma_5 \\ \sigma_6 \end{bmatrix} = [T] \begin{bmatrix} \sigma_x \\ \sigma_y \\ \sigma_z \\ \sigma_{yz} \\ \sigma_{xz} \\ \sigma_{xy} \end{bmatrix} \quad \text{and} \quad \begin{bmatrix} \epsilon_1 \\ \epsilon_2 \\ \epsilon_3 \\ \epsilon_4 \\ \epsilon_5 \\ \epsilon_6 \end{bmatrix} = [T] \begin{bmatrix} \epsilon_x \\ \epsilon_y \\ \epsilon_z \\ \epsilon_{yz} \\ \epsilon_{xz} \\ \epsilon_{xy} \end{bmatrix} \tag{2.28}
$$

wherein

$$
[T] = \begin{bmatrix} m^2 & n^2 & 0 & 0 & 0 & 2mn \\ n^2 & m^2 & 0 & 0 & 0 & -2mn \\ 0 & 0 & 1 & 0 & 0 & 0 \\ 0 & 0 & 0 & m & -n & 0 \\ 0 & 0 & 0 & n & m & 0 \\ -mn & mn & 0 & 0 & 0 & (m^2 - n^2) \end{bmatrix}. \tag{2.29}
$$

For completeness

$$
\begin{bmatrix} \sigma_x \\ \sigma_y \\ \sigma_z \\ \sigma_{yz} \\ \sigma_{xz} \\ \sigma_{xy} \end{bmatrix} = [T]^{-1} \begin{bmatrix} \sigma_1 \\ \sigma_2 \\ \sigma_3 \\ \sigma_4 \\ \sigma_5 \\ \sigma_6 \end{bmatrix} \quad \text{and} \quad \begin{bmatrix} \epsilon_x \\ \epsilon_y \\ \epsilon_z \\ \epsilon_{yz} \\ \epsilon_{xz} \\ \epsilon_{xy} \end{bmatrix} = [T]^{-1} \begin{bmatrix} \epsilon_1 \\ \epsilon_2 \\ \epsilon_3 \\ \epsilon_4 \\ \epsilon_5 \\ \epsilon_6 \end{bmatrix} \tag{2.30}
$$

where *

$$
[T]^{-1} = \begin{bmatrix} m^2 & n^2 & 0 & 0 & 0 & -2mn \\ n^2 & m^2 & 0 & 0 & 0 & 2mn \\ 0 & 0 & 0 & 0 & 0 & 0 \\ 0 & 0 & 0 & m & n & 0 \\ 0 & 0 & 0 & -n & m & 0 \\ mn & -mn & 0 & 0 & 0 & (m^2 - n^2) \end{bmatrix}. \tag{2.31}
$$

* $[T]^{-1}$ can be found by replacing θ by $(-\theta)$ in $[T]$

If one systematically uses these expressions, and utilizes Hooke's Law relating stress and strain, and includes the thermal and hygrothermal effects, one can produce the following overall general equations for a lamina of a fiber reinforced composite material in terms of the principal material directions (1, 2, 3); see equations (2.5) through (2.8).

$$
\begin{Bmatrix} \sigma_1 \\ \sigma_2 \\ \sigma_3 \\ \sigma_4 \\ \sigma_5 \\ \sigma_6 \end{Bmatrix} = \begin{bmatrix} Q_{11} & Q_{12} & Q_{13} & 0 & 0 & 0 \\ Q_{12} & Q_{22} & Q_{23} & 0 & 0 & 0 \\ Q_{13} & Q_{23} & Q_{33} & 0 & 0 & 0 \\ 0 & 0 & 0 & 2Q_{44} & 0 & 0 \\ 0 & 0 & 0 & 0 & 2Q_{55} & 0 \\ 0 & 0 & 0 & 0 & 0 & 2Q_{66} \end{bmatrix} \begin{Bmatrix} \epsilon_1 - \alpha_1 \Delta T - \beta_1 \Delta m \\ \epsilon_2 - \alpha_2 \Delta T - \beta_2 \Delta m \\ \epsilon_3 - \alpha_3 \Delta T - \beta_3 \Delta m \\ \epsilon_{23} \\ \epsilon_{31} \\ \epsilon_{12} \end{Bmatrix} .
$$

$$(2.32)$$

In the above, the Q_{ij} quantities are used for the stiffness matrix quantitites because modern composite materials technology uses them in all literature, but they are identical to the C_{ij} quantities of classical elasticity discussed in earlier sections, and can be obtained directly from (2.5) through (2.17). One should also remember that $\epsilon_{23} = (1/2G_{23})\sigma_4$, $\epsilon_{31} = (1/2G_{31})\sigma_5$ and $\epsilon_{12} = (1/2G_{12})\sigma_6$, hence the coefficients of "two" appearing in the Q matrix. Using the notation of Sloan [15]

$$
\begin{aligned}
&Q_{11} = E_{11}(1 - \nu_{23}\nu_{32})/\Delta, \quad Q_{22} = E_{22}(1 - \nu_{31}\nu_{13})/\Delta \\
&Q_{33} = E_{33}(1 - \nu_{12}\nu_{21})/\Delta, \quad Q_{44} = G_{23}, \quad Q_{55} = G_{13}, \quad Q_{66} = G_{12} \\
&Q_{12} = (\nu_{21} + \nu_{31}\nu_{23})E_{11}/\Delta = (\nu_{12} + \nu_{32}\nu_{13})E_{22}/\Delta \\
&Q_{13} = (\nu_{31} + \nu_{21}\nu_{32})E_{11}/\Delta = (\nu_{13} + \nu_{12}\nu_{23})E_{22}/\Delta \\
&Q_{23} = (\nu_{32} + \nu_{12}\nu_{31})E_{22}/\Delta = (\nu_{23} + \nu_{21}\nu_{13})E_{33}/\Delta \\
&\Delta = 1 - \nu_{12}\nu_{21} - \nu_{23}\nu_{32} - \nu_{31}\nu_{13} - 2\nu_{21}\nu_{32}\nu_{13}
\end{aligned}
$$

$$(2.33)$$

Incidentally in the above expressions, if the lamina is transversely isotropic, i.e., same properties in both the 2 and 3 direction, then $\nu_{12} = \nu_{13}$, $G_{12} = G_{13}$, $E_{22} = E_{33}$ with resulting simplification.

For preliminary calculation in design or where great accuracy is not needed, simpler forms [6] for some of the expressions in (2.33) can be used, as shown below:

$$Q_{11} = E_{11}/(1 - \nu_{12}\nu_{21}), \quad Q_{22} = E_{22}/(1 - \nu_{12}\nu_{21})$$

$$Q_{12} = Q_{21} - \nu_{21}E_{11}/(1 - \nu_{12}\nu_{21}) = \nu_{12}E_{22}/(1 - \nu_{12}\nu_{21}) \qquad (2.34)$$

$$Q_{66} = G_{12}$$

If these simpler forms are used then one would use the classical form of the constitutive relations instead of (2.32), neglecting transverse shear deformation and transverse normal stress, namely

$$\begin{Bmatrix} \sigma_1 \\ \sigma_2 \\ \sigma_6 \end{Bmatrix} = \begin{bmatrix} Q_{11} & Q_{12} & 0 \\ Q_{12} & Q_{22} & 0 \\ 0 & 0 & 2Q_{66} \end{bmatrix} \begin{Bmatrix} \epsilon_1 - \alpha_1 \Delta T - \beta_1 \Delta m \\ \epsilon_2 - \alpha_2 \Delta T - \beta_2 \Delta m \\ \epsilon_{12} \end{Bmatrix} \tag{2.35}$$

where one should remember also that $2\epsilon_{12} = \epsilon_6$, hence the appearance of two before Q_{66}.

Now, to relate these relationships to the x-y-z coordinate system, one utilizes equations (2.30) through (2.32). The result is

$$\begin{Bmatrix} \sigma_x \\ \sigma_y \\ \sigma_z \\ \sigma_{yz} \\ \sigma_{xz} \\ \sigma_{xy} \end{Bmatrix} = \begin{bmatrix} \overline{Q}_{11} & \overline{Q}_{12} & \overline{Q}_{13} & 0 & 0 & 2\overline{Q}_{16} \\ \overline{Q}_{12} & \overline{Q}_{22} & \overline{Q}_{23} & 0 & 0 & 2\overline{Q}_{26} \\ \overline{Q}_{13} & \overline{Q}_{23} & \overline{Q}_{33} & 0 & 0 & 2\overline{Q}_{36} \\ 0 & 0 & 0 & 2\overline{Q}_{44} & 2\overline{Q}_{45} & 0 \\ 0 & 0 & 0 & 2\overline{Q}_{45} & 2\overline{Q}_{55} & 0 \\ \overline{Q}_{16} & \overline{Q}_{26} & \overline{Q}_{36} & 0 & 0 & 2\overline{Q}_{66} \end{bmatrix} \begin{Bmatrix} \epsilon_x - \alpha_x \Delta T - \beta_x \Delta m \\ \epsilon_y - \alpha_y \Delta T - \beta_y \Delta m \\ \epsilon_z - \alpha_z \Delta T - \beta_z \Delta m \\ \epsilon_{yz} \\ \epsilon_{xz} \\ \epsilon_{xy} - \frac{1}{2}\alpha_{xy} \Delta T - \frac{1}{2}\beta_{xy} \Delta m \end{Bmatrix} \tag{2.36}$$

where $[\overline{Q}] = [T]^{-1}[Q][T]$, or more explicitly,

$$\overline{Q}_{11} = Q_{11}m^4 + 2(Q_{12} + 2Q_{66})m^2n^2 + Q_{22}n^4$$

$$\overline{Q}_{12} = (Q_{11} + Q_{22} - 4Q_{66})m^2n^2 + Q_{12}(m^4 + n^4)$$

$$\overline{Q}_{13} = Q_{13}m^2 + Q_{23}n^2$$

$$\overline{Q}_{16} = -mn^3Q_{22} + m^3nQ_{11} - mn(m^2 - n^2)(Q_{12} + 2Q_{66})$$

$$\overline{Q}_{22} = Q_{11}n^4 + 2(Q_{12} + 2Q_{66})m^2n^2 + Q_{22}m^4$$

$$\overline{Q}_{23} = n^2Q_{13} + m^2Q_{23}$$

$$\overline{Q}_{33} = Q_{33}$$

$$\overline{Q}_{26} = -m^3nQ_{22} + mn^3Q_{11} + mn(m^2 - n^2)(Q_{12} + 2Q_{66})$$

$$\overline{Q}_{36} = (Q_{13} - Q_{23})mn$$

$$\overline{Q}_{44} = Q_{44}m^2 + Q_{55}n^2$$

$$\overline{Q}_{45} = (Q_{55} - Q_{44})mn$$

$$\overline{Q}_{55} = Q_{55}m^2 + Q_{44}n^2$$

$$\overline{Q}_{66} = (Q_{11} + Q_{22} - 2Q_{12})m^2n^2 + Q_{66}(m^2 - n^2)^2$$

$$\alpha_x = \alpha_1 m^2 + \alpha_2 n^2 \qquad \beta_x = \beta_1 m^2 + \beta_2 n^2$$
$$\alpha_y = \alpha_2 m^2 + \alpha_1 n^2 \qquad \beta_y = \beta_2 m^2 + \beta_1 n^2$$
$$\alpha_z = \alpha_3 \qquad\qquad \beta_z = \beta_3$$
$$\alpha_{xy} = (\alpha_1 - \alpha_2) mn \qquad \beta_{xy} = (\beta_1 - \beta_2) mn.$$

It should be remembered that although the coefficients of both thermal and hygrothermal expansion are purely dilatational in the material coordinate system 1-2, rotation into the structural coordinate system x-y, results in an α_{xy} and a β_{xy}.

Again, for preliminary design purposes or for approximate calculations one can use the simpler classical form of

$$\begin{Bmatrix} \sigma_x \\ \sigma_y \\ \sigma_{xy} \end{Bmatrix} = \begin{bmatrix} \overline{Q}_{11} & \overline{Q}_{12} & 2\overline{Q}_{16} \\ \overline{Q}_{12} & \overline{Q}_{22} & 2\overline{Q}_{26} \\ \overline{Q}_{16} & \overline{Q}_{26} & 2\overline{Q}_{66} \end{bmatrix} \begin{Bmatrix} \epsilon_x - \alpha_x \Delta T - \beta_x \Delta m \\ \epsilon_y - \alpha_y T - \beta_y \Delta m \\ \epsilon_{xy} - \frac{1}{2}\alpha_{xy}\Delta T - \frac{1}{2}\beta_{xy}\Delta m \end{Bmatrix} \qquad (2.37)$$

where the $\overline{Q}_{ij}$ are defined in (2.36), but one must use the Q_{ij} of (2.34) instead of (2.33) for consistency.

One interesting variation of the above classical quantities of (2.37) resulted when Pagano and Chou rewrote many of the above quantities in a manner that is very useful in comparing various material systems in the design of a composite structure.

$$\overline{Q}_{11} = U_1 + U_2 \cos(2\theta) + U_3 \cos(4\theta)$$

$$\overline{Q}_{22} = U_1 - U_2 \cos(2\theta) + U_2 \cos(4\theta)$$

$$\overline{Q}_{12} = U_4 - U_3 \cos(4\theta)$$

$$\overline{Q}_{66} = U_5 - U_3 \cos(4\theta)$$

$$\overline{Q}_{16} = +\tfrac{1}{2}U_2 \sin(2\theta) + U_3 \sin(4\theta)$$

$$\overline{Q}_{26} = +\tfrac{1}{2}U_2 \sin(2\theta) - U_3 \sin(4\theta)$$

where

$$U_1 = \tfrac{1}{8}(3Q_{11} + 3Q_{22} + 2Q_{12} + 4Q_{66})$$

$$U_2 = \tfrac{1}{2}(Q_{11} - Q_{22})$$

$$U_3 = \tfrac{1}{8}(Q_{11} + Q_{22} - 2Q_{12} - 4Q_{66})$$

$$U_4 = \tfrac{1}{8}(Q_{11} + Q_{22} + 6Q_{12} - 4Q_{66})$$

$$U_5 = \tfrac{1}{8}(Q_{11} + Q_{22} - 2Q_{12} + 4Q_{66})$$

In the above, the U_i quantities are invariant with respect to axis rotation and therefore are truly composite lamina properties.

At this point, given a lamina of a unidirectional composite of known elastic properties, if used in a plate or panel, with the 1-2 material axis at an angle θ from the plate or panel x-y axes, all stiffness quantities Q_{ij} and $\overline{Q}_{ij}$ can be determined relating stresses and strains in either coordinate system.

2.6. Laminate Analysis

In the previous section the generalized constitutive equations for one lamina of a composite material were formulated. In reality, any structure of composite materials is comprised of numerous laminae which are bonded and/or cured together. In fact, over and above the superior properties in strength and stiffness that composites possess, the ability to stack laminae one on the other in a varied but unique fashion to result in the optimum laminate material properties for a given structural size and set of loadings is one of the biggest advantages that composites have over metallic or plastic structures.

Consider a laminate composed of N laminae. For the kth lamina of the laminate, equation (2.36) can be written

$$
\begin{Bmatrix} \sigma_x \\ \sigma_y \\ \sigma_z \\ \sigma_{yz} \\ \sigma_{xz} \\ \sigma_{xy} \end{Bmatrix}_k = [\overline{Q}]_k \begin{Bmatrix} \epsilon_x - \alpha_x \Delta T - \beta_x \Delta m \\ \epsilon_y - \alpha_y \Delta T - \beta_y \Delta m \\ \epsilon_z - \alpha_z \Delta T - \beta_z \Delta m \\ \epsilon_{yz} \\ \epsilon_{xz} \\ \epsilon_{xy} - \frac{1}{2}\alpha_{xy} \Delta T - \frac{1}{2}\beta_{xy} \Delta m \end{Bmatrix}_k \tag{2.38}
$$

where all matrices must have the subscript k due to the orientation of the particular lamina with respect to the plate or shell x-y coordinates and its unique $\overline{Q}$, α_i and β_i.

For any elastic body the strain-displacement equations, i.e., those kinematic relations describing the functional relations between the elastic strains in the body and its displacements, are given by

$$
\epsilon_{ij} = \frac{1}{2}(u_{i,j} + u_{j,i}) \tag{2.39}
$$

where i, $j = x$, y, z in a Cartesian coordinate frame, and the comma denotes partial differentiation with respect to the coordinate denoted by the symbol after the comma. Explicitly, the relations are:

$$\epsilon_x = \frac{\partial u}{\partial x}, \quad \epsilon_y = \frac{\partial v}{\partial y}, \quad \epsilon_z = \frac{\partial w}{\partial z}$$

$$\left. \begin{array}{c} \epsilon_{xz} = \frac{1}{2}\left(\frac{\partial u}{\partial z} + \frac{\partial w}{\partial x}\right), \quad \epsilon_{yz} = \frac{1}{2}\left(\frac{\partial v}{\partial z} + \frac{\partial w}{\partial y}\right) \\[3mm] \epsilon_{xy} = \frac{1}{2}\left(\frac{\partial u}{\partial y} + \frac{\partial v}{\partial x}\right). \end{array} \right\} \qquad (2.40)$$

In the above u, v, and w are the displacements in the x, y, and z directions respectively. In linear elastic plate theory [2], it is assumed that a lineal element extending through the thickness of a thin plate and perpendicular to the middle surface (that is, the x-y plane in Figure 2.9 below) prior to loading, upon the load application undergoes at most a translation and a rotation with respect to the original coordinate system. Based upon that one assumption the functional form of the displacements for the plate are:

$$u(x, y, z) = u_0(x, y) + z\bar{\alpha}(x, y)$$

$$v(x, y, z) = v_0(x, y) + z\bar{\beta}(x, y) \qquad (2.41)$$

$$w(x, y) = w(x, y)$$

where u_0, v_0 and w_0 are the middle surface displacements, i.e., the translations of the lineal element, and the second terms in the first two equations are related to the rotations. In classical beam and plate theory $\bar{\alpha}$ and $\bar{\beta}$ would be the negative of the first derivative of the lateral displacement with respect to the x and y coordinates respectively (i.e., $\bar{\alpha} = -(\partial w / \partial x)$ and $\bar{\beta} = -(\partial w / \partial y)$, the negative of the slope), but if transverse shear deformation of the plate is non-zero, $\bar{\alpha}$ and $\bar{\beta}$ take on another relationship from the above, which will be discussed later. Also,

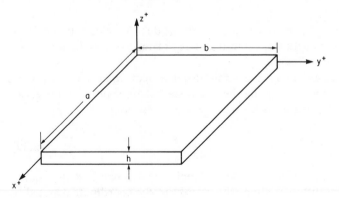

Figure 2.9. Typical Rectangular Plate.

in classical plate theory, it is assumed that the lineal element across the thickness of the plate cannot extend or shrink because at most it undergoes a translation and rotation, hence $w = w(x, y)$ only. Sloan [15] has shown that for all practical composites, plate thickening, i.e., $w(x, y, z)$ is unimportant and can be neglected, hence it is not included here, but can be studied in the reference cited.

Substituting (2.41) into (2.39) results in:

$$\epsilon_x = \frac{\partial u_0}{\partial x} + z \frac{\partial \bar{\alpha}}{\partial x}, \quad \epsilon_y = \frac{\partial v_0}{\partial y} + z \frac{\partial \bar{\beta}}{\partial y}, \quad \epsilon_z = 0$$

$$\epsilon_{xz} = \frac{1}{2}\left(\bar{\alpha} + \frac{\partial w}{\partial x} \right), \quad \epsilon_{yz} = \frac{1}{2}\left(\bar{\beta} + \frac{\partial w}{\partial y} \right) \tag{2.42}$$

$$\epsilon_{xy} = \frac{1}{2}\left(\frac{\partial u_0}{\partial y} + \frac{\partial v_0}{\partial x} \right) + \frac{z}{2}\left(\frac{\partial \bar{\alpha}}{\partial y} + \frac{\partial \bar{\beta}}{\partial x} \right).$$

The mid surface strains can be written as:

$$\epsilon_{x_0} = \frac{\partial u_0}{\partial x}, \quad \epsilon_{y_0} = \frac{\partial v_0}{\partial y}, \quad \epsilon_{xy_0} = \frac{1}{2}\left(\frac{\partial u_0}{\partial y} + \frac{\partial v_0}{\partial x} \right). \tag{2.43}$$

The curvatures can be written as

$$\kappa_x = \frac{\partial \bar{\alpha}}{\partial x}, \quad \kappa_y = \frac{\partial \bar{\beta}}{\partial y}, \quad \kappa_{xy} = \frac{1}{2}\left(\frac{\partial \bar{\alpha}}{\partial y} + \frac{\partial \bar{\beta}}{\partial x} \right). \tag{2.44}$$

The plate theory assumed displacements and strains are employed to describe the strains and displacements of a laminated plate of composite materials, because all laminae are bonded together such that the same assumptions are made regarding the lineal element through the thickness. Such continuity of strains and displacements occurs regardless of the orientation of individual plies or laminae.

Substituting (2.42) into (2.38) results in the following, wherein because we have assumed $\epsilon_z = 0$, because plate thickening is neglected, σ_z for a thin-walled structure of composite material is usually negligible and will not be considered further.

$$\begin{Bmatrix} \sigma_x \\ \sigma_y \\ \sigma_{yz} \\ \sigma_{xz} \\ \sigma_{xy} \end{Bmatrix}_k = [\bar{Q}]_k \begin{Bmatrix} \epsilon_{x_0} + z\kappa_x - \alpha_x \Delta T - \beta_x \Delta m \\ \epsilon_{y_0} + z\kappa_y - \alpha_y \Delta T - \beta_y \Delta m \\ \epsilon_{yz} \\ \epsilon_{xz} \\ \epsilon_{xy_0} + \kappa_{xy} z - \frac{1}{2}\alpha_{xy}\Delta T - \frac{1}{2}\beta_{xy}\Delta m \end{Bmatrix}_k \tag{2.45}$$

Note that without the hygrothermal terms, the strain-curvature matrix would suffice for the entire laminate independent of orientation, because the displacements, and strains are continuous over the thickness of the laminate. In that case the subscript k on that matrix would not be needed. However, even though there is continuity of the mid-surface strains and curvatures across the thickness of the laminate, the stresses are discontinuous across the laminate thickness because of the various orientations of the laminae, hence, the subscript k in the stress matrix above.

It is seen from (2.45) that if all quantities on the right hand side are known, one can easily calculate each stress component in each lamina comprising the laminate.

Consider a laminated plate or panel of thickness h as shown below, in Figure 2.10.

It is seen that h_k is the *vectorial* distance from the panel mid-plane, $z = 0$, to the upper surface of the kth lamina, i.e., any dimension below the midsurface is a negative dimension and any dimension above the midsurface is positive. For example, consider a laminate 0.52 mm (0.020″) thick, composed of four equally thick laminae, each being 0.13 mm (0.005″) thick. Then $h_0 = -0.25$ mm $(-0.010″)$, $h_1 = -0.13$ mm $(-0.005″)$, $h_2 = 0$, $h_3 = 0.13$ mm (0.005″) and $h_4 = 0.25$ mm (0.010″).

As in classical beam, plate and shell theory [1–6], one defines and uses stress resultants (N), stress couples (M), and shear resultants (Q) for the overall plate regardless of the number and the orientation of the laminae, hence:

$$\begin{Bmatrix} N_x \\ N_y \\ N_{xy} \\ Q_x \\ Q_y \end{Bmatrix} = \int_{-h/2}^{+h/2} \begin{Bmatrix} \sigma_x \\ \sigma_y \\ \sigma_{xy} \\ \sigma_{xz} \\ \sigma_{yz} \end{Bmatrix} dz, \quad \begin{Bmatrix} M_x \\ M_y \\ M_{xy} \end{Bmatrix} = \int_{-h/2}^{+h/2} \begin{Bmatrix} \sigma_x \\ \sigma_y \\ \sigma_{xy} \end{Bmatrix} z \, dz. \quad (2.46)$$

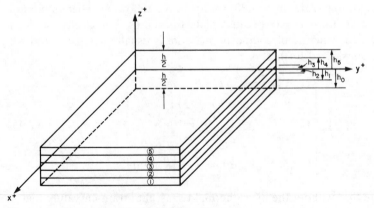

Figure 2.10. Nomenclature for the Stacking Sequence.

On the plate shown in Figure 2.11, the positive directions of all the resultants, and stress couples are shown, consistent with the definitions of the quantities given in (2.46).

For a laminated plate, the stress components can be integrated across each lamina, but must then be added together as follows; employing equations (2.38) and (2.42) through (2.44)

$$
\begin{bmatrix} N_x \\ N_y \\ N_{xy} \end{bmatrix} = \sum_{k=1}^{N} \int_{h_{k-1}}^{h_k} \begin{bmatrix} \sigma_x \\ \sigma_y \\ \sigma_{xy} \end{bmatrix} dz
$$

$$
= \sum_{k=1}^{N} \left\{ \int_{h_{k-1}}^{h_k} [\overline{Q}]_k \begin{bmatrix} \epsilon_{x0} \\ \epsilon_{y0} \\ \epsilon_{xy_0} \end{bmatrix} dz + \int_{h_{k-1}}^{h_k} [\overline{Q}]_k \begin{bmatrix} \kappa_x \\ \kappa_y \\ \kappa_{xy} \end{bmatrix} z dz \right.
$$

$$
\left. - \int_{h_{k-1}}^{h_k} [\overline{Q}]_k \begin{bmatrix} \alpha_x \\ \alpha_y \\ \frac{1}{2}\alpha_{xy} \end{bmatrix}_k \Delta T dz - \int_{h_{k-1}}^{h_k} [\overline{Q}]_k \begin{bmatrix} \beta_x \\ \beta_y \\ \frac{1}{2}\beta_{xy} \end{bmatrix} \Delta m dz \right\} \quad (2.47)
$$

where here only the pertinent portions of the $[Q]_k$ matrix are used.

Since the derivatives of the mid-surface displacements (u_0 and v_0) and the rotations ($\bar{\alpha}$ and $\bar{\beta}$) and the $\overline{Q}$'s are not functions of z, (2.47) can be rewritten as:

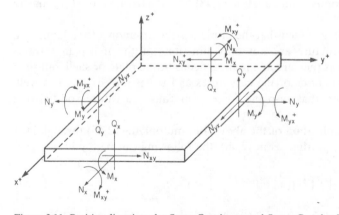

Figure 2.11. Positive directions for Stress Resultants and Stress Couples for a Plate.

$$
\begin{bmatrix} N_x \\ N_y \\ N_{xy} \end{bmatrix} = \sum_{k=1}^{N} \left\{ [\overline{Q}]_k \begin{bmatrix} \epsilon_{x0} \\ \epsilon_{y0} \\ \epsilon_{xy0} \end{bmatrix} \int_{h_{k-1}}^{h_k} dz + [\overline{Q}]_k \begin{bmatrix} \kappa_x \\ \kappa_y \\ \kappa_{xy} \end{bmatrix} \int_{h_{k-1}}^{h_k} z \, dz \right.
$$

$$
\left. - \int_{h_{k-1}}^{h_k} [\overline{Q}]_k \begin{bmatrix} \alpha_x \\ \alpha_y \\ \frac{1}{2}\alpha_{xy} \end{bmatrix} \Delta T dz - \int_{h_{k-1}}^{h_k} [\overline{Q}]_k \begin{bmatrix} \beta_x \\ \beta_y \\ \frac{1}{2}\beta_{xy} \end{bmatrix} \Delta m dz \right\}. \quad (2.48)
$$

Finally, (2.48) can be written as:

$$
[N] = [A][\epsilon_0] + [B][\kappa] - [N]^T - [N]^m, \quad (2.49)
$$

where it is shown later in (2.62) that a factor of 2 is necessary in some terms, and where

$$
A_{ij} = \sum_{k=1}^{N} (\overline{Q}_{ij})_k [h_k - h_{k-1}], \qquad [i, j = 1, 2, 6] \quad (2.50)
$$

$$
B_{ij} = \frac{1}{2} \sum_{k=1}^{N} (\overline{Q}_{ij})_k [h_k^2 - h_{k-1}^2], \qquad [i, j = 1, 2, 6] \quad (2.51)
$$

$$
N_{ij}^T = \sum_{k=1}^{N} \int_{h_{k-1}}^{h_k} (\overline{Q}_{ij})_k [\alpha_{ij}]_k \Delta T dz, \qquad [i, j = 1, 2, 6] \quad (2.52)
$$

$$
N_{ij}^m = \sum_{k=1}^{N} \int_{h_{k-1}}^{h_k} (\overline{Q}_{ij})_k [\beta_{ij}]_k \Delta m dz, \qquad [i, j = 1, 2, 6] \quad (2.53)
$$

where it is obvious from equation (2.48) how the $[\alpha_{ij}]_k$ and $[\beta_{ij}]_k$ matrix are defined.

From (2.49), it is seen that the in-plane stress resultants for a laminated thin-walled structure are not only functions of the mid-plane strains ($\partial u_0/\partial x$, etc.) as they are in a homogeneous beam, plate or shell, but they can also be functions of the curvatures and twists ($\partial \bar{\alpha}/\partial x$, etc.) as well. Also, it is seen that in-plane forces can cause curvatures or twisting deformations.

By exact duplication of the above, but multiplying (2.45) through by z first before integrating, as in (2.46), the following can be found:

$$
[M] = [B][\epsilon_0] + [D][\kappa] - [M]^T - [M]^m \quad (2.54)
$$

where it will be shown in (2.62) that factors of 2 are necessary in some terms, and where

$$D_{ij} = \frac{1}{3} \sum_{k=1}^{N} (\overline{Q}_{ij})_k [h_k^3 - h_{k-1}^3] \tag{2.55}$$

$$M_{ij}^T = \sum_{k=1}^{N} \int_{h_{k-1}}^{h_k} (\overline{Q}_{ij})_k [\alpha_{ij}]_k \Delta T z \, dz \tag{2.56}$$

$$M_{ij}^m = \sum_{k=1}^{N} \int_{h_{k-1}}^{h_k} (\overline{Q}_{ij})_k [\beta_{ij}]_k \Delta m z \, dz \tag{2.57}$$

where $i, j = 1, 2, 6$

To determine the shear resultants Q_x and Q_y, defined in equation (2.46), it is assumed that the transverse shear stresses are distributed parabolically across the laminate thickness. In spite of the discontinuities at the interface between laminae, a continuous function $f(z)$ is used as a weighting function, which is consistent with Reissner:

$$f(z) = \frac{5}{4}\left[1 - \left(\frac{z}{h/2}\right)^2\right]. \tag{2.58}$$

Then from (2.36), (2.42), (2.46), and (2.58)

$$\sigma_{xz_k} = 2\overline{Q}_{55_k}\epsilon_{xz} + 2\overline{Q}_{45_k}\epsilon_{yz}$$

$$\sigma_{yz_k} = 2\overline{Q}_{45_k}\epsilon_{xz} + 2\overline{Q}_{44_k}\epsilon_{yz}$$

hence

$$Q_x = 2(A_{55}\epsilon_{xz} + A_{45}\epsilon_{yz}) \tag{2.59}$$

$$Q_y = 2(A_{45}\epsilon_{xz} + A_{44}\epsilon_{yz}) \tag{2.60}$$

where

$$A_{ij} = \frac{5}{4} \sum_{k=1}^{N} (\overline{Q}_{ij})_k \left[h_k - h_{k-1} - \frac{4}{3}(h_k^3 - h_{k-1}^3)\frac{1}{h^2}\right] \tag{2.61}$$

where $h = $ (thickness of laminate) [2], and $i, j = 4, 5$ only.

Finally, the results of (2.49) and (2.54) can be written succinctly as follows:

$$
\begin{bmatrix} N_x \\ N_y \\ N_{xy} \\ \hline M_x \\ M_y \\ M_{xy} \end{bmatrix} = \begin{bmatrix} A_{11} & A_{12} & 2A_{16} & B_{11} & B_{12} & 2B_{16} \\ A_{12} & A_{22} & 2A_{26} & B_{12} & B_{22} & 2B_{26} \\ A_{16} & A_{26} & 2A_{66} & B_{16} & B_{26} & 2B_{66} \\ \hline B_{11} & B_{12} & 2B_{16} & D_{11} & D_{12} & 2D_{16} \\ B_{12} & B_{22} & 2B_{26} & D_{12} & D_{22} & 2D_{26} \\ B_{16} & B_{26} & 2B_{66} & D_{16} & D_{26} & 2D_{66} \end{bmatrix} \begin{bmatrix} \epsilon_{x_0} \\ \epsilon_{y_0} \\ \epsilon_{xy_0} \\ \hline \kappa_x \\ \kappa_y \\ \kappa_{xy} \end{bmatrix} - \begin{bmatrix} N_x^T \\ N_y^T \\ N_{xy}^T \\ \hline M_x^T \\ M_y^T \\ M_{xy}^T \end{bmatrix} - \begin{bmatrix} N_x^m \\ N_y^m \\ N_{xy}^m \\ \hline M_x^m \\ M_y^m \\ M_{xy}^m \end{bmatrix} \quad (2.62)
$$

It is seen that the $[A]$ matrix is the extensional stiffness matrix relating the in-plane stress resultants (N's) to the mid-surface strains (ϵ_0's) and the $[D]$ matrix is the flexural stiffness matrix relating the stress couples (M's) to the curvatures (κ's). Since the $[B]$ matrix relates M's to ϵ_0's and N's to κ's, it is called the bending-stretching coupling matrix. It should be noted that a laminated structure can have bending-stretching coupling even if all lamina are isotropic, for example, a laminate composed of one lamina of steel and another of polyester. In fact, only when the structure is exactly symmetric about its middle surface are all of the B_{ij} components equal to zero, and this requires symmetry in laminae properties, orientation, and location from the middle surface.

It is seen that stretching-shearing coupling occurs when A_{16} and A_{26} are non-zero. Twisting-stretching coupling as well as bending-shearing coupling occurs when the B_{16} and B_{26} terms are non-zero, and bending-twisting coupling comes from non-zero values of the D_{16} and D_{26} terms. Usually the 16 and 26 terms are avoided by proper stacking sequences, but there could be some structural applications where these effects could be used to advantage, such as in aeroelastic tailoring.

Examples of these effects in several cross-ply laminates (i.e., combinations of $0°$ and $90°$ plies), angle ply laminates (combinations of $+\theta$ and $-\theta$ plies), and unidirectional laminates (all $0°$ plies) are involved in problems at the end of this chapter. It is seen in (2.50) that $(h_k - h_{k-1})$ is always positive, and from (2.55) $(h_k^3 - h_{k-1}^3)$ is always positive, hence, in all cross-ply laminates, all ( )$_{16}$ and ( )$_{26}$ terms are zero.

At this point the stress strain relations, or constitutive relations (2.62) can be combined with the appropriate stress equations of equilibrium, and the strain-displacement relations to form an appropriate beam, plate or shell theory including thermal and hygrothermal effects as well as transverse shear deformation. These are discussed in detail by Vinson and Chou,[6] and herein in Chapters 3, 4 and 5.

2.7. Problems

1. Consider a laminate composed of boron-epoxy with the following

properties:

$$Q_{11} = 2.43 \times 10^5 \text{ MPA} \qquad U_1 = 1.074 \times 10^5 \text{ MPA}$$
$$(35.32 \times 10^6 \text{ psi}) \qquad (15.58 \times 10^6 \text{ psi})$$
$$Q_{22} = 2.43 \times 10^4 \text{ MPA} \qquad U_2 = 1.096 \times 10^5 \text{ MPA}$$
$$(3.532 \times 10^6 \text{ psi}) \qquad (15.9 \times 10^6 \text{ psi})$$
$$Q_{12} = 7.30 \times 10^3 \text{ MPA} \qquad U_3 = 2.646 \times 10^4 \text{ MPA}$$
$$(1.06 \times 10^6 \text{ psi}) \qquad (3.84 \times 10^6 \text{ psi})$$
$$Q_{66} = 1.034 \times 10^4 \text{ MPA} \qquad U_4 = 3.121 \times 10^4 \text{ MPA}$$
$$(1.5 \times 10^6 \text{ psi}) \qquad (4.53 \times 10^6 \text{ psi})$$
$$U_5 = 3.417 \times 10^4 \text{ MPA}$$
$$(4.96 \times 10^6 \text{ psi})$$

If the laminate is a cross-ply with [0°/90°/90°/0°], with each ply being 0.25 mm (0.01″) thick, and if the laminate is loaded in tension in the x direction (i.e., the 0° direction):

(a) What percentage of the load is carried by the 0° plies? the 90° plies?

(b) If the strength of the 0° plies is 1.364×10^3 MPA (198,000 psi), and the strength of the 90° plies is 44.8 MPA (6,500 psi) which plies will fail first?

(c) What is the maximum load, $N_{x_{max}}$, that the laminate can carry at incipient failure? What stress exists in the remaining two plies, at the failure load of the two others?

(d) If the structure can tolerate failure of two plies, what is the maximum load, $N_{x_{max}}$ that the other two plies can withstand to failure?

2. A laminate is composed of graphite epoxy (GY70/339) with the following properties: $E_{11} = 2.89 \times 10^5$ MPA (42×10^6 psi), $E_{22} = 6.063 \times 10^3$ MPA (0.88×10^6 psi), $G_{12} = 4.134 \times 10^3$ MPA (0.6×10^6 psi), and $\nu_{12} = 0.31$. Determine the elements of the A, B and D matrices for a two-ply laminate [+45°/−45°], where each ply is 0.15 mm (0.006″) thick.

3. Consider a *square* panel composed of *one* ply with the fibers shown in the directions shown in Figure 2.12.

Which of the orientations above would be the stiffest for the loads given in Figure 2.13.

4. For a panel consisting of boron-epoxy with the properties of Problem 1 above, and a stacking sequence of [0°/+45°/−45°/0°], and a ply thickness of 0.15 mm (0.006″), determine the elements of the A, B and D matrices. What would the elements be if the ply thickness was 0.14 mm (0.0055″)?

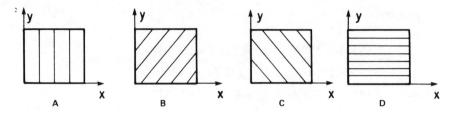

Figure 2.12.

5. The properties of graphite fibers and a polyimide matrix are as follows:

$E = 2.756 \times 10^5$ MPA $E = 2.756 \times 10^3$ MPA
$\left(40 \times 10^6 \text{ psi}\right)$ $\left(0.4 \times 10^6 \text{ psi}\right)$
$\nu = 0.2$ $\nu = 0.33$

(a) Find the modulus of elasticity in the fiber direction, E_{11}, of a laminate of graphite-polyimide composite with 60% fiber volume ratio.
(b) Find the Poisson's ratio, ν_{12}.
(c) Find the modulus of elasticity normal to the fiber direction, E_{22}.
(d) What is the Poisson's ratio, ν_{21}?
   6. Consider a laminate composed of GY70/339 graphite epoxy whose properties are given above in Problem 2. For a lamina thickness of 0.127 mm (0.005″), calculate the elements of the A, B and D matrices for the following laminates:
   (a) unidirectional, 4 plies
   (b) [0°, 90°, 90°, 0°]

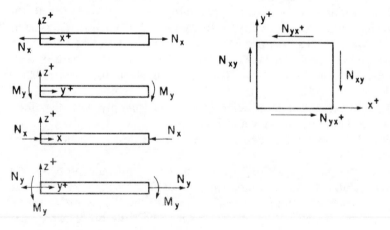

Figure 2.13.

(c) $[\pm 45]_s$, i.e., $[+45°/-45°/-45°/+45°]$

(d) $[0°/+45°/-45°/90°]_s$

7. Consider a laminate composed of GY70/339 graphite epoxy whose properties are given in Problem 2. Using the lamina thickness 0.127 mm (0.005″) cited in Problem 6, calculate the elements of the A, B, and D matrices for the following laminates:

(a) $[\pm(45)_2]_s$, $[+45/-45/+45/-45]_s$

(b) $[\pm 45]_s$, $[+45/-45/-45/+45]$

(c) $[\pm 45]_{Qs}$, $[+45/-45/+45/-45]$

(d) $[\pm(45)_2]_{Qs}$, $[+45/-45/+45/-45]_{Qs}$

Compare the forms of the A, B, and D matrices between laminate types.

8. What type of coupling would you expect in the (B) matrix for (a), (b): (that is, identify the non-zero terms)

(a) $0°/90°$ laminate

(b) $+\theta/-\theta$ laminate

9. Given a composite laminate composed of continuous fiber laminae of High Strength Graphite/Epoxy with properties of Table 2.2, if the laminate architecture is $[0°, 90°, 90°, 0°]$, determine A_{11} if $\nu_{12} = 0.3$, and each ply thickness is 0.006″.

10. Consider a plate composed of a 0.01″ thick steel plate joined perfectly to an aluminum plate, 0.01″ thick. Using the properties of Table 2.2 calculate B_{11}, if the Poisson's Ratio of each materiial is $\nu = 0.3$.

11. Consider a unidirectional composite composed of a polyimide matrix and graphite fibers with properties given in Problem 5. In the fiber direction, what volume fraction is required to have a composite stiffness of $E_{11} = 10 \times 10^6$ psi to match an aluminum stiffness.

12. A laminate is composed of ultra high modulus graphite epoxy with properties given in Table 2.2. Determine the elements of the A, B and D matrices for a two ply laminate $[+45°/-45°]$, where each ply is 0.006″ thick.

13. For a panel consisting of boron-epoxy with the properties of Problem 2.1 and a stacking sequence of $[0/+45°/-45°/0°]$, and a ply thickness of 0.006″, determine the elements of the A, B and D matrices.

14. Consider a composite lamina made up of continuous Boron fibers imbedded in an epoxy matrix. The diameter of the Boron fibers is 0.004 in and the volume fraction of Boron fibers in the composite is 40%. Assuming that the modulus of elasticity of the Boron fibers is 5×10^7 psi and the epoxy is 5×10^5 psi, find:

(a) The Young's moduli of the composite in the 1 and 2 directions.

(b) Consider an identical second lamina to be glued to the first so that the fibers of the second lamina are parallel to the 2 direction. Assuming the thickness of each lamina to be 0.1″ and neglecting Poisson's ratio, what are the new moduli in the 1 and 2 directions?

15. The properties of graphite fibers and a polyimide matrix are as

Table 2.2 Unidirectional properties

Material	Elastic moduli			Ultimate strength			Density
	Axial E_{11}	Transverse E_{22}	Shear G_{12}	Axial tens. σ_{11}	Trans. tens. σ_{22}	shear τ_{12}	
High strength	20	1.0	0.65	220	6	14	0.057
GR/epoxy	(138)	(6.9)	(4.5)	(1517)	(41)	(97)	(1.57)
High modulus	32	1.0	0.7	175	5	10	0.058
GR/epoxy	(221)	(6.9)	(4.8)	(1206)	(34)	(69)	(1.60)
Ultra high	44	1.0	0.95	110	4	7	0.061
modulus	(303)	(6.9)	(6.6)	(758)	(28)	(48)	(1.68)
GR/epoxy							
Kevlar 49/	12.5	0.8	0.3	220	4	6	0.050
epoxy	(86)	(5.5)	(2.1)	(1517)	(28)	(41)	(1.38)
S glass	8	1.0	0.5	260	6	10	0.073
epoxy	(55)	(14)	(3.4)	(1793)	(41)	(69)	(2.00)
Steel	30	30	11.5	60	60	35	0.284
	(207)	(207)	(79)	(414)	(414)	(241)	(7.83)
Aluminum	10.5	10.5	3.8	42	42	28	0.098
6061-T6	(72)	(72)	(26)	(290)	(290)	(193)	(2.70)

Moduli in Msi (GPa)
Stress in Ksi (kPa)
Density in lb/in³ (g/cm³)

follows

GRAPHITE   POLYIMIDE

$E = 40 \times 10^6$ psi $E = 0.4 \times 10^6$ psi

$\nu = 0.2$ $\nu = 0.33$.

(a) Find the modulus of elasticity in the fiber direction, E_{11}, of a lamina of graphite – polyimide composite with 70% fiber volume ratio.
(b) Find the Poisson's ratio, ν_{12}.
(c) Find the modulus of elasticity normal to the fiber direction, E_{22}.
(d) What is the Poisson's ratio ν_{21}?

16. In a given composite, the coefficient of thermal expansion for the epoxy and the graphite fibers are $+30 \times 10^{-6}$ in/in/°F and -15×10^{-6} in/in/°F respectively. For space application where no thermal distortion can be tolerated what volume fractions of each component are required to make zero expansion and contraction in the fiber direction for an all 0° construction? (Hint: Use the Rule of Mixtures)

17. Find the A, B, and D matrices for the following composite: *50% Volume Fraction Boron-Epoxy Composite*

$E_{11} = 30.0 \times 10^6$

$E_{22} = 3.0 \times 10^6$

$G_{12} = 1.0 \times 10^6$

$\nu_{12} = 0.22$

Stacking Sequence (each lamina is 0.0125″ thick)

$\theta = 45°$
$\theta = 0°$
$\theta = 90°$
$\theta = -45°$
$\theta = -45°$
$\theta = 90°$
$\theta = 0°$
$\theta = 45°$

18. Three composite plates are under uniform transverse loading. All the conditions, such as materials, boundary conditions and geometry, etc. are the same except the stacking sequence as shown below. Without using any calculation, indicate which plate will have maximum deflection and which will have minimum deflection.

$[45/-45/45/-45]s$, $[45/45/-45/-45]s$, $[0/0/45/45]s$

2.8. References

1. Shames, I.H. *Introduction to Solid Mechanics.* New York: Prentice-Hall, Inc., 1975.
2. Vinson, J.R. *Structural Mechanics: The Behavior of Plates and Shells.* New York: Wiley-Interscience, John Wiley and Sons, Inc., 1974.
3. Timoshenko, S.A., and Woinowsky-Krieger. *Theory of Plates and Shells.* New York: McGraw-Hill Book Company, Inc., 1959.

62

4. Marguerre, H., and Woernle, H.T. *Elastic Plates.* New York: Blaisdell Publishing Company, 1969.
5. Sokolnikoff, I.S. *Mathematical Theory of Plasticity.* New York: MacGraw-Hill Book Company, 1956.
6. Vinson, J.R., and Chou, T.W. *Composite Materials and Their Use in Structures.* London: Applied Science Publishers, 1975.
7. Halpin, J.C., and Tsai, S.W. "Environmental Factors in Composite Materials Design," Air Force Materials Laboratory Technical Report 67-423, 1967.
8. Hashin, Z. "Theory of Fiber Reinforced Materials," National Aeronautics and Space Administration Contractors Report 1974 (1972).
9. Christensen, R.M. *Mechanics of Composite Materials.* New York: John Wiley and Sons, Inc., 1979.
10. Hahn, T.H. "Simplified Formulas for Elastic Moduli of Unidirectional Continuous Fiber Composites," *Composites Technology Review,* Fall 1980.
11. Vinson, J.R., Walker, W.J., Pipes, R.B., and Ulrich, O.R. "The Effects of Relative Humidity and Elevated Temperatures on Composite Structures," Air Force Office of Scientific Research Technical Report 77-0030, February 1977.
12. Pipes, R.B., Vinson, J.R., and Chou, T.W. "On the Hygrothermal Response of Laminated Composite Systems," *Journal of Composite Materials,* April 1976: 130–148.
13. Flaggs, D.L., and Vinson, J.R. "Hygrothermal Effects on the Buckling of Laminated Composite Plates," *Fibre Science and Technology,* Vol. 11, No. 5, September 1978: 353–365.
14. Flaggs, D.L. "Elastic Stability of Generally Laminated Composite Plates Including Hygrothermal Effects", MMAE Thesis, University of Delaware, 1978.
15. Sloan, J.G. "The Behavior of Rectangular Composite Material Plates Under Lateral and Hygrothermal Loads", MMAE Thesis, University of Delaware, 1979.

3. Plates and Panels of Composite Materials

3.1. Introduction

In Chapter 2, the governing constitutive equations were developed in detail, describing the relationships between integrated stress resultants (N_x, N_y, N_{xy}), integrated stress couples (M_x, M_y, M_{xy}), in-plane mid-surface strains $(\epsilon_x^0, \epsilon_y^0, \epsilon_{xy}^0)$, and the curvatures $(\kappa_x, \kappa_y, \kappa_{xy})$, as seen in Equations (2.62). These will be utilized with the strain-displacement relations of (2.40) and (2.42) and the equilibrium equations to be developed in Section 3.2 to develop structural theories for thin walled bodies, the structural form in which composite materials are most generally employed. Plates and panels are discussed in this chapter. Beams, rods and columns are discussed in Chapter 4. Shells will be the subject of Chapter 5. The use of energy methods for solving structures problems is discussed in Chapter 6. However to study any of the structural equations it is necessary to develop suitable equilibrium equations first.

3.2. Plate Equilibrium Equations

The integrated stress resultants (N), shear resultants (Q) and stress couples (M) are defined by (2.46), and their positive directions are shown in Figure 2.11, for a rectangular plate, defined as a body of length a in the x-direction, width b in the y-direction, and thickness h in the z-direction, where $h \ll b$, $h < a$, that is, a thin plate. It is now necessary to derive the equilibrium relations which exist between them.

In mathematically modeling solid materials, including the laminates of Chapter 2, a continuum theory is generally employed. In doing so, a representative material point within the elastic solid or lamina is selected as being typical of all material points in the body or lamina. The material point is assumed to be infinitely smaller than any dimension of the structure containing it, but infinitely larger than the size of the lattice spacing of the structured material comprising it. Moreover, the material point is given a convenient shape; and in a Cartesian reference frame that convenient shape is a small cube of dimensions dx, dy and dz as shown in Figure 3.1 below.

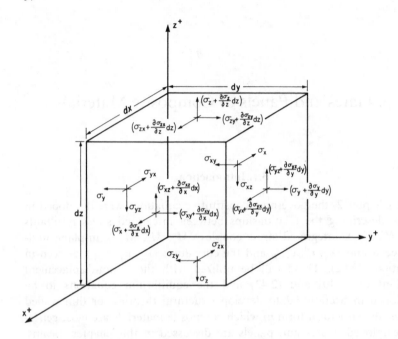

Figure 3.1. Coordinate systems, geometry, and nomenclature.

This cubic material point of dimensions dx, dy and dz is termed a control element. The positive values of all stresses acting on each surface of the control element are shown in Figure 3.1, along with how they vary from one surface to another, using the positive sign convention consistent with most scientific literature, and consistent with Figure 2.1. Details of the nomenclature can be found in any text on solid mechanics, including Vinson [1], and Vinson and Chou [2]. In addition to the surface stresses acting on the control element shown in Figure 3.1, body force components F_x, F_y and F_z can also act on the body. These force components are proportional to the control element volume (i.e., its mass), such as gravitational, magnetic or centrifugal forces for example.

A force balance can now be made in the x, y and z directions resulting in three equations of equilibrium. For instance, a force balance on the x-direction would yield

$$\left(\sigma_x + \frac{\partial \sigma_x}{\partial x}dx\right)dydz + \left(\sigma_{yx} + \frac{\partial \sigma_{yx}}{\partial y}dy\right)dxdz + \left(\sigma_{zx} + \frac{\partial \sigma_{zx}}{\partial z}dz\right)dxdy$$

$$-\sigma_x dydz - \sigma_{yx}dxdz + \sigma_{zx}dydx + F_x dxdydz = 0$$

Cancelling certain terms and dividing the remaining terms by the volume results in the following

$$\frac{\partial \sigma_x}{\partial x} + \frac{\partial \sigma_{yx}}{\partial y} + \frac{\partial \sigma_{zx}}{\partial z} + F_x = 0. \tag{3.1}$$

Similarly, equilibrium in the y and z directions yields:

$$\frac{\partial \sigma_{xy}}{\partial x} + \frac{\partial \sigma_y}{\partial y} + \frac{\partial \sigma_{zy}}{\partial z} + F_y = 0 \tag{3.2}$$

$$\frac{\partial \sigma_{xz}}{\partial x} + \frac{\partial \sigma_{yz}}{\partial y} + \frac{\partial \sigma_z}{\partial z} + F_z = 0 \tag{3.3}$$

These three equations comprise the equilibrium equations for three dimensional elasticity. However, for beam, plate and shell theory, whether involving composite materials or not, one must integrate the stresses across the thickness of the thin walled structures to obtain solutions.

Recalling the definitions of Chapter 2 for the stress resultants and stress couples:

$$\begin{bmatrix} N_x \\ N_y \\ N_{xy} \\ Q_x \\ Q_y \end{bmatrix} = \int_{-h/2}^{+h/2} \begin{bmatrix} \sigma_x \\ \sigma_y \\ \sigma_{xy} \\ \sigma_{xz} \\ \sigma_{yz} \end{bmatrix} dz = \sum_{k=1}^{N} \int_{h_{k-1}}^{h_k} \begin{bmatrix} \sigma_x \\ \sigma_y \\ \sigma_{xy} \\ \sigma_{xz} \\ \sigma_{yz} \end{bmatrix}_k dz_k \tag{3.4}$$

$$\begin{bmatrix} M_x \\ M_y \\ M_{xy} \end{bmatrix} = \int_{-h/2}^{h/2} \begin{bmatrix} \sigma_x \\ \sigma_y \\ \sigma_{xy} \end{bmatrix} z dz = \sum_{k=1}^{N} \int_{h_{k-1}}^{h_k} \begin{bmatrix} \sigma_x \\ \sigma_y \\ \sigma_{xy} \end{bmatrix}_k z_k dz_k \tag{3.5}$$

The first form of each is applicable to a single layer plate, while the second form is necessary for a laminated plate due to the stress discontinuities associated with different materials and/or differing orientations in the various plies.

Turning now to (3.1), neglecting the body force term, F_x, for simplicity of example, integrating term by term across each ply, and summing across the plate provides

$$\sum_{k=1}^{N} \int_{h_{k-1}}^{h_k} \frac{\partial \sigma_{x_k}}{\partial x} dz + \sum_{k=1}^{N} \int_{h_{k-1}}^{h_k} \frac{\partial \sigma_{yx_k}}{\partial y} dz + \sum_{k=1}^{N} \int_{h_{k-1}}^{h_k} \frac{\partial \sigma_{zx_t}}{\partial z} dz = 0 \tag{3.6}$$

In the first two terms integration and differentiation can be interchanged, hence:

$$\frac{\partial}{\partial x}\left[\sum_{k=1}^{N}\int_{h_{k-1}}^{h_k}\sigma_{x_k}dz\right]+\frac{\partial}{\partial y}\left[\sum_{k=1}^{N}\int_{h_{k-1}}^{h_k}\sigma_{yx_k}dz\right]+\sum_{k=1}^{N}\sigma_{zx}\bigg]_{h_{k-1}}^{h_k}=0 \qquad (3.7)$$

In the first two terms N_x and N_{yx} appear explicitly. In the third term it is clear that between all plies the interlaminar shear stresses will cancel each other out, and if one defines applied surface shear stresses on the top $(z = h_N)$ and bottom $(z = h_0)$ surfaces as defined below (see Figure 2.1):

$$\sigma_{zx}(h_N)\equiv\tau_{1x}\quad\text{and}\quad\sigma_{zx}(h_0)\equiv\tau_{2x} \qquad (3.8)$$

Equation (3.7) can be written as:

$$\frac{\partial N_x}{\partial x}+\frac{\partial N_{yx}}{\partial y}+\tau_{1x}-\tau_{2x}=0 \qquad (3.9)$$

Similarly, integrating the equilibrium equation in the y-direction provides

$$\frac{\partial N_{xy}}{\partial x}+\frac{\partial N_y}{\partial y}+\tau_{1y}-\tau_{2y}=0 \qquad (3.10)$$

where

$$\sigma_{zy}(h_N)\equiv\tau_{1y}\quad\text{and}\quad\sigma_{zy}(h_0)\equiv\tau_{2y} \qquad (3.11)$$

Likewise equilibrium in the z-direction upon integration and summing provides

$$\frac{\partial Q_x}{\partial x}+\frac{\partial Q_y}{\partial y}+p_1-p_2=0 \qquad (3.12)$$

where

$$\sigma_z(h_N)\equiv p_1\quad\text{and}\quad\sigma_z(h_0)\equiv p_2 \qquad (3.13)$$

In addition to the integrated force equilibrium equations above, two equations of moment equilibrium are also needed, one for the x-direction and one for the y-direction. Multiplying equation 3.1 through by zdz, integrating across each ply and summing across all laminae results in the following:

$$\sum_{k=1}^{N} \int_{h_{k-1}}^{h_k} \frac{\partial \sigma_{x_k}}{\partial x} z \, dz + \sum_{k=1}^{N} \int_{h_{k-1}}^{h_k} \frac{\partial \sigma_{yx_k}}{\partial y} z \, dz + \sum_{k=1}^{N} \int_{h_{k-1}}^{h_k} \frac{\partial \sigma_{zx_k}}{\partial z} z \, dz = 0$$

Again, in the first two terms integration and summation can be interchanged with differentiation with the result that the first two terms become $(\partial M_x/\partial x) + (\partial M_{xy}/\partial y)$. However, the third term must be integrated by parts as follows:

$$\sum_{k=1}^{N} \int_{h_{k-1}}^{h_k} \frac{\partial \sigma_{zx_k}}{\partial z} z \, dz = \sum_{k=1}^{N} \left\{ z\sigma_{zx}]_{h_{k-1}}^{h_k} - \int_{h_{k-1}}^{h_k} \sigma_{zx} dz \right\}$$

Here the last term is clearly $-Q_x$. Again in the first term on the right, clearly the moments of all the interlaminar stresses between plies cancel each other out, and the only non-zero terms are the moments of the applied surface shear stresses hence that term becomes

$$\frac{h}{2} [\tau_{1x} + \tau_{2x}] = h_N \tau_{1x} - h_0 \tau_{2x}$$

Using the former expression, the equation of equilibrium of moments in the x-direction is

$$\frac{\partial M_x}{\partial x} + \frac{\partial M_{xy}}{\partial y} - Q_x + \frac{h}{2} [\tau_{1x} + \tau_{2x}] = 0 \qquad (3.14)$$

Similarly in the y-direction the moment equilibrium equation is

$$\frac{\partial M_{xy}}{\partial x} + \frac{\partial M_y}{\partial y} - Q_y + \frac{h}{2} [\tau_{1y} + \tau_{2y}] = 0, \qquad (3.15)$$

where all the terms are defined above. Thus, there are five equilibrium equations for a rectangular plate, regardless of what material or materials are utilized in the plate: (3.9), (3.10), (3.12), (3.14) and (3.15).

3.3. The Bending of Composite Material Plates

Consider a plate composed of a laminated composite material that is mid-plane symmetric, that is $B_{ij} = 0$, and has no other coupling terms, $(\ )_{16} = (\ )_{26} = 0$, no surface shear stresses and no hygrothermal effects.

The plate equilibrium equations for the bending of the plate, due to lateral loads given by (3.14), (3.15) and (3.12) become:

$$\frac{\partial M_x}{\partial x} + \frac{\partial M_{xy}}{\partial y} - Q_x = 0 \tag{3.16}$$

$$\frac{\partial M_{xy}}{\partial x} + \frac{\partial M_y}{\partial y} - Q_y = 0 \tag{3.17}$$

$$\frac{\partial Q_x}{\partial x} + \frac{\partial Q_y}{\partial y} + p(x, y) = 0 \tag{3.18}$$

Equations, (3.16) and (3.17) can be substituted into (3.18) with the result that:

$$\frac{\partial^2 M_x}{\partial x^2} + 2\frac{\partial^2 M_{xy}}{\partial x \partial y} + \frac{\partial^2 M_y}{\partial y^2} = -p(x, y) \tag{3.19}$$

where $p(x, y) = p_1(x, y) - p_2(x, y)$.

The above equations are derived from equilibrium considerations alone. From (2.62) and for the case of mid-plane symmetry ($B_{ij} = 0$) and no ( )$_{16}$ and ( )$_{26}$ terms, the constitutive relations are:

$$M_x = D_{11}\kappa_x + D_{12}\kappa_y \tag{3.20}$$

$$M_y = D_{12}\kappa_x + D_{22}\kappa_y \tag{3.21}$$

$$M_{xy} = 2D_{66}\kappa_{xy} \tag{3.22}$$

where from (2.44)

$$\kappa_x = \frac{\partial \bar{\alpha}}{\partial x}, \quad \kappa_y = \frac{\partial \bar{\beta}}{\partial y}, \quad \kappa_{xy} = \frac{1}{2}\left(\frac{\partial \bar{\alpha}}{\partial y} + \frac{\partial \bar{\beta}}{\partial x}\right)$$

It is well known that transverse shear deformation (that is $\epsilon_{xz} \neq 0$, $\epsilon_{yz} \neq 0$) effects are important in plates composed of composite materials in determining maximum deflections, vibration natural frequencies and critical buckling loads, but a simpler stress analysis for preliminary design, which excludes these effects to determine a "first cut" for stresses, the required overall stacking sequence and required plate thickness is appropriate.

If, in fact, transverse shear deformation is ignored, then from (2.42)

$$\epsilon_{xz} = 0 = \frac{1}{2}\left(\bar{\alpha} + \frac{\partial w}{\partial x}\right), \quad \text{and} \quad \epsilon_{yz} = 0 = \frac{1}{2}\left(\bar{\beta} + \frac{\partial w}{\partial y}\right),$$

hence,

$$\bar{\alpha} = -\frac{\partial w}{\partial x} \quad \text{and} \quad \bar{\beta} = -\frac{\partial w}{\partial y}, \quad \text{and} \quad \kappa_x = -\frac{\partial^2 w}{\partial x^2}, \quad \kappa_y = -\frac{\partial^2 w}{\partial y^2}$$

$$\text{and} \quad \kappa_{xy} = -\frac{\partial^2 w}{\partial x \partial y} \tag{3.23}$$

So, substituting (3.23) into (3.20) through (3.22) results in:

$$M_x = -D_{11}\frac{\partial^2 w}{\partial x^2} - D_{12}\frac{\partial^2 w}{\partial y^2} \tag{3.24}$$

$$M_y = -D_{12}\frac{\partial^2 w}{\partial x^2} - D_{22}\frac{\partial^2 w}{\partial y^2} \tag{3.25}$$

$$M_{xy} = -2D_{66}\frac{\partial^2 w}{\partial x \partial y} \tag{3.26}$$

Substituting these three equation in turn into (3.19) results in:

$$D_{11}\frac{\partial^4 w}{\partial x^4} + 2(D_{12} + 2D_{66})\frac{\partial^4 w}{\partial x^2 \partial y^2} + D_{22}\frac{\partial^4 w}{\partial y^4} = p(x, y) \tag{3.27}$$

The above coefficients can be simplified to:

$$D_{11} \equiv D_1, \quad D_{22} \equiv D_2, \quad (D_{12} + 2D_{66}) \equiv D_3 \tag{3.28}$$

with the result that (3.27) becomes

$$D_1\frac{\partial^4 w}{\partial x^4} + 2D_3\frac{\partial^4 w}{\partial x^2 \partial y^2} + D_2\frac{\partial^4 w}{\partial y^4} = p(x, y) \tag{3.29}$$

This is the governing differential equation for the bending of a plate composed of a composite material, excluding transverse shear deformation, no coupling terms (that is $B_{ij} = (\ )_{16} = (\ )_{26} = 0$), and no hygrothermal terms (that is, $\Delta T = \Delta m = 0$) subjected to a lateral distributed load $p(x, y)$. As stated previously, neglecting transverse shear deformation and hygrothermal effects can lead to significant errors, as will be shown, but in many cases their neglect results in easier solutions which are useful in preliminary design to "size" the plate initially.

Solutions of (3.29) can be obtained generally in two ways: direct solution of the governing differential equation (3.29), or utilization of an energy principle solution. The latter offers more latitude in finding approximate solutions, often needed, and these will be discussed in detail later in Chapter 6.

Direct solutions of the governing differential equations for plates of composite materials fall into three categories: Navier solutions, Levy solutions and perturbation solutions. Each has its advantages and disadvantages. However, prior to that boundary conditions need to be discussed.

3.4. Plate Boundary Conditions

In "classical" (that is, ignoring transverse shear deformation) plate theory of Section 3.3, two and only two boundary conditions are needed at each edge of the plate. The boundary conditions for a simply supported edge and a clamped edge shown below are identical to those of classical beam theory, where n is the direction normal to the plate edge and t is the direction parallel or tangent to the edge:

Simply Supported Edge

$$w = 0 \quad \text{and} \quad M_n = 0 \tag{3.30}$$

$M_n = 0$ implies $\partial^2 w / \partial n^2 = 0$, because in (3.24) and (3.25), there is no curvature (that is $\partial^2 w / \partial t^2$) along the edge of the simply supported plate because $w = 0$.

Clamped Edge

$$w = 0 \quad \text{and} \quad \frac{\partial w}{\partial n} = 0 \tag{3.31}$$

Free Edge

For a free edge of a plate, that is, one on which there are no loads nor any displacement or slope requirements, it is clear that M_n, Q_n and M_{nt} all should be zero. However with classical plate theory only two boundary conditions can be satisfied. This was a major problem for the solid mechanists of the early nineteenth century. Then Kirchoff brilliantly formulated an approximate solution to the problem, which is developed in many texts [1,2]. He reasoned that for the free edge

$$M_n = 0 \tag{3.32}$$

where now because there is curvature along the edge the full expressions of (3.24) and (3.25) must be utilized. The approximate expression for the second boundary condition is

$$V_n = Q_n + \frac{\partial M_{nt}}{\partial t} = 0 \tag{3.33}$$

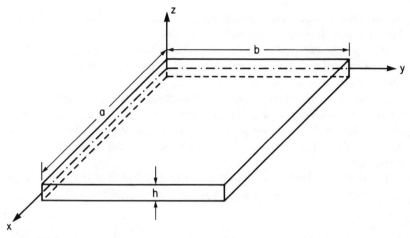

Figure 3.2.

where V_n is the "effective" shear resultant, and Q_n is given by (3.16) or (3.17), M_n is given by (3.24) or (3.25), and M_{nt} is given by (3.26).

3.5. Navier Solutions for Plates of Composite Materials

Direct solutions for the bending of rectangular plates as the one shown in Figure 3.2 of composite materials, that is, the solution to (3.29), can be classified into three approaches. In the Navier approach for the case of the plate being simply supported on all four edges, one simply expands the lateral deflection, $w(x, y)$ and the applied lateral load, $p(x, y)$, into a doubly infinite half range sine series because that series satisfies all of the boundary conditions (3.30) exactly:

$$w(x, y) = \sum_{m=1}^{\infty} \sum_{n=1}^{\infty} A_{mn} \sin\frac{m\pi x}{a} \sin\frac{n\pi y}{b} \qquad (3.34)$$

$$p(x, y) = \sum_{m=1}^{\infty} \sum_{n=1}^{\infty} B_{mn} \sin\frac{m\pi x}{a} \sin\frac{n\pi y}{b} \qquad (3.35)$$

3.6. Navier Solution for a Uniformly Loaded Simply Supported Plate

The case of a uniformly loaded, $p(x, y) = -p_0$ simply supported plate is solved by means of the Navier series solution of Section 3.5, for two composite materials systems: unidirectional and cross-ply laminates. The in-plane stresses σ_x, σ_y and σ_{xy} are determined for each case at the quarter points and mid-point of the plate. In addition, the solutions have been examined by utilizing one, three and five terms in the Navier series solution.

In this analysis, it is of course assumed that all plies are perfectly bonded and classical theory is used (that is, ϵ_{xz} and ϵ_{yz} are assumed zero). This results in the in-plane stresses, σ_x, σ_y and σ_{xy} being directly determined, while the transverse shear stresses σ_{xz} and σ_{yz} are determined subsequently.

Using the methods discussed previously it is found that the stresses in each lamina for the case of $p(x, y) = -p_0$ are given by:

$$
\begin{Bmatrix} \sigma_x \\ \sigma_y \\ \sigma_{xy} \end{Bmatrix}_k = + \frac{16 p_{0z}}{\pi^3} \sum_{m=1,3,5}^{\infty} \sum_{n=1,3,5}^{\infty} \frac{1}{mnD}
$$

$$
\cdot \begin{bmatrix} \left[-Q_{11}^k \left(\frac{m}{a}\right)^2 - Q_{12}^k \left(\frac{n}{b}\right)^2 \right] \sin\frac{m\pi x}{a} \cdot \sin\frac{n\pi y}{b} \\ \left[-Q_{12}^k \left(\frac{m}{a}\right)^2 - Q_{22}^k \left(\frac{n}{b}\right)^2 \right] \sin\frac{m\pi x}{a} \cdot \sin\frac{n\pi y}{b} \\ 2Q_{66}^k \left(\frac{m}{a}\right)\left(\frac{n}{b}\right) \cos\frac{m\pi x}{a} \cos\frac{n\pi y}{b} \end{bmatrix} \tag{3.36}
$$

where

$$
D = D_{11}\left(\frac{m}{a}\right)^4 + 2(D_{12} + 2D_{66})\left(\frac{mn}{ab}\right)^2 + D_{22}\left(\frac{n}{b}\right)^4
$$

In this numerical example a square plate of $a = b = 12''$ is considered. The total plate thickness is $0.08''$: light plies of $0.01''$ thickness ($h_k = 0.01''$).

The first material system considered is E glass/epoxy, with the following properties:

$E_{11} = 8.8 \times 10^6$ psi

$E_{22} = 3.6 \times 10^6$ psi

$\nu_{12} = 0.23$

$G_{12} = 1.74 \times 10^6$ psi

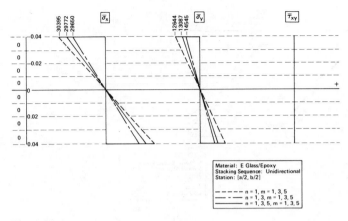

Figure 3.3.

For a fiber volume fraction of 0.70, the stiffnesses Q_{ij} are

0° ply (psi)	90° ply (psi)
$Q_{11} = 9.0 \times 10^6$	$Q_{11} = 3.68 \times 10^6$
$Q_{12} = 0.85 \times 10^6$	$Q_{12} = 0.85 \times 10^6$
$Q_{22} = 3.68 \times 10^6$	$Q_{22} = 9.0 \times 10^6$
$Q_{66} = 1.74 \times 10^6$	$Q_{66} = 1.74 \times 10^6$

In the following figures, the stresses have been normalized as $\bar{\sigma}_{ij} = \sigma_{ij}/p_0$. In Figure 3.3, the normalized stresses are shown at plate midpoint ($x = a/2$, $y = b/2$) for a unidirectional ($\theta = 0°$) laminate. In Figure 3.4, the normalized stresses at plate midpoint are shown for a mid-plane symmetric cross-ply laminate. In Figure 3.5, these stresses are shown at the quarter point location for the cross-ply plate.

A cross-ply laminate of T300-5208 graphite-epoxy has been used for comparison with the E glass/epoxy laminate. Properties of the graphite-epoxy laminate are:

$$E_{11} = 22.2 \times 10^6 \text{ psi}$$
$$E_{22} = 1.58 \times 10^6 \text{ psi} \qquad V_f = 70\%$$
$$\nu = 0.12$$
$$G_{12} = 0.81 \times 10^6 \text{ psi}$$

Again normalized stresses have been shown in Figures 3.6 and 3.7 at both plate quarter point and midpoint.

74

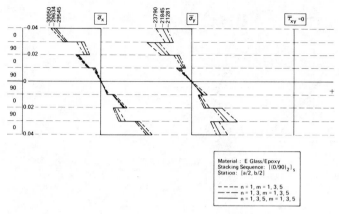

Figure 3.4.

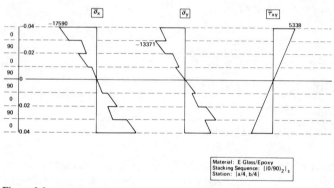

Figure 3.5.

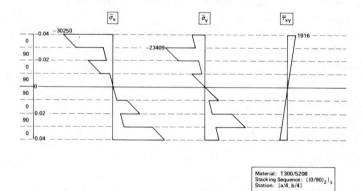

Figure 3.6.

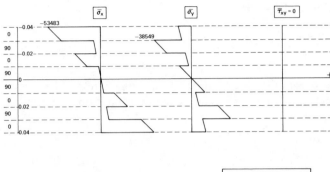

Figure 3.7.

Some conclusions can be drawn from this example set.

1. Solution convergence is rapid within the framework of taking three terms for both m and n for evaluating $\bar{\sigma}_x$, but is not as rapid in calculating $\bar{\sigma}_y$.

2. For the same material there is little difference between the maximum value of the $\bar{\sigma}_x$ stress for both the unidirectional and cross-ply composites at similar plate locations, however, the $\bar{\sigma}_y$ stresses differ significantly.

3. The stress $\bar{\sigma}_y$ at a fixed location for the graphite/epoxy laminate is much smaller relative to the $\bar{\sigma}_x$ value (10%), compared to that in the E glass/epoxy laminate where $\bar{\sigma}_y$ is 33% of the value of $\bar{\sigma}_x$ at the same location.

This example was performed by Wenn-Jinn Liou, a student at the University of Florida.

3.7. Levy Solution for Plates of Composite Materials

The second direct method of solution for the bending of rectangular plates due to lateral loads, is due to Maurice Levy [3], who in 1899 introduced a single infinite series method of solution for isotropic plate problems. The method can also be used to solve (3.29) for a composite material plate.

Consider a composite material plate, shown in Figure 3.2, with edges $y = 0$ and $y = b$ simply supported. The boundary conditions on those edges are

$$w(x, 0) = w(x, b) = 0$$
$$M_y(x, 0) = M_y(x, b) = 0 \tag{3.37}$$

The latter implies that

$$\frac{\partial^2 w(x, 0)}{\partial y^2} = \frac{\partial^2 w(x, b)}{\partial y^2} = 0 \tag{3.38}$$

Levy assumed the following solution form: a single infinite half range sine series which satisfies the boundary conditions on both y edges:

$$w(x, y) = \sum_{n=1}^{\infty} \phi_n(x) \sin \frac{n\pi y}{b} \tag{3.39}$$

where $\phi_n(x)$ is at this point an unknown function of x. A laterally distributed load $p(x, y)$ can be expressed as follows:

$$p(x, y) = g(x)h(y) \tag{3.40}$$

where $g(x)$ and $h(y)$ are specified. The form of (3.39) requires that the $h(y)$ portion of the load be expanded also in terms of a half range sine series also, such as

$$h(y) = \sum_{h=1}^{\infty} A_n \sin \frac{n\pi y}{b} \tag{3.41}$$

where

$$A_n = \frac{2}{b} \int_0^b h(y) \sin \frac{n\pi y}{b} dy \tag{3.42}$$

Substituting (3.39) through (3.41) into (3.29), and observing that the equation exists only if it is true term by term, it is seen that, after dividing by D_1:

$$\phi_n^{IV}(x) - \frac{2D_3}{D_1} \lambda_n^2 \phi_n''(x) + \frac{D_2}{D_1} \lambda_n^4 \phi_n(x) = \frac{A_n g_n(x)}{D_1} \tag{3.43}$$

where $\lambda n = n\pi/b$.

Note that (3.43) was derived without specifying any boundary conditions on the x edges. In fact, the homogeneous solution of (3.43) yields four constants of integration which are determined through satisfying the boundary conditions on those edges.

To obtain the homogeneous solution of (3.43), the right hand side is equal to zero:

$$\phi_n^{IV}(x) - \frac{2D_3}{D_1} \lambda_n^2 \phi_n''(x) + \frac{D_2}{D_1} \lambda_n^4 \phi_n(x) = 0 \tag{3.44}$$

After letting $\phi_n(x) = e^{sx}$, and dividing the result by e^{sx}, (3.44) becomes:

$$s^4 - \frac{2D_3}{D_1}\lambda_n^2 s^2 + \frac{D_2}{D_1}\lambda_n^4 = 0 \tag{3.45}$$

Unlike the case of an isotropic plate where $D_1 = D_2 = D_3$, such that the roots s are easily seen to be $\pm\lambda_n$ and $\pm\lambda_n$ (repeated roots), for this case there are three sets of roots depending upon whether $(D_2/D_1)^{1/2}$ is greater than, equal to or less than D_3/D_1. Hence, for the composite plate using the Levy type solution three different forms for the homogeneous solution of $\phi_n(x)$ to be put into (3.39) exist depending on the relative plate stiffness in various directions.

For the case, $(D_2/D_1)^{1/2} < (D_3/D_1)$

$$\phi_{n_H}(x) = C_1 \cosh(\lambda_n s_1 x) + C_2 \sinh(\lambda_n s_1 x)$$

$$+ C_3 \cosh(\lambda_n s_2 x) + C_4 \sinh(\lambda_n s_2 x) \tag{3.46}$$

where the roots are:

$$s_1 = \sqrt{\left(\frac{D_3}{D_1}\right) + \sqrt{\left(\frac{D_3}{D_1}\right)^2 - \frac{D_2}{D_1}}}, \quad s_2 \sqrt{\frac{D_3}{D_1} - \sqrt{\left(\frac{D_3}{D_1}\right)^3 - \frac{D_2}{D_1}}}$$

For the case, $(D_2/D_1)^{1/2} = (D_3/D_1)$

$$\phi_{n_H}(x) = (C_5 + C_6 x)\cosh(\lambda_n s_3 x)$$

$$+ (C_7 + C_8 x)\sinh(\lambda_n s_3 x) \tag{3.47}$$

where the roots are

$$s_3 = \pm\sqrt{\frac{D_3}{D_1}}$$

For the case $(D_2/D_1)^{1/2} > (D_3/D_1)$

$$\phi_{n_H}(x) = (C_9 \cos\lambda_n s_5 x + C_{10} \sin\lambda_n s_5 x)\cosh(\lambda_n s_4 x)$$

$$+ (C_{11} \cos\lambda_n s_5 x + C_{12} \sin\lambda_n s_5 x)\sinh(\lambda_n s_4 x) \tag{3.48}$$

where the roots are

$$s_4 = \sqrt{\frac{1}{2}\left[\left(\frac{D_2}{D_1}\right)^{1/2} + \frac{D_3}{D_1}\right]}, \quad s_5 = \sqrt{\frac{1}{2}\left[\left(\frac{D_2}{D_1}\right)^{1/2} - \frac{D_3}{D_1}\right]}$$

Obviously, for a given plate whose materials and orientation have already been specified (the analysis problem) only one of the three cases exists. However, if one is trying to determine the best material and orientation (the design problem) then more than one case may be involved, with the necessity of determining not just four constants, but eight or all twelve to satisfy the edge boundary conditions to determine which construction is best for the design.

Concerning the particular solution, it is noted that if the lateral load, $p(x, y)$, is at most linear in x, hence from (3.40), $g_n(x)$ is at most linear in x, then from (3.43) the particular solution is:

$$\phi_{n_p}(x) = \frac{A_n g_n(x)}{\lambda_n^4 D_2}$$

Otherwise, one must seek a particular solution. In any case, one must then add the relevant ϕ_{n_H} to the proper ϕ_{n_p}, in order to satisfy any set of boundary conditions on the x edges of the plate. For example, suppose the $x = 0$ edge is simply supported, then from (3.30) the boundary conditions are:

$$w(0, y) = 0, \quad \text{and} \quad M_x(0, y) = 0 \Rightarrow \frac{\partial^2 w}{\partial x^2}(0, y) = 0 \tag{3.49}$$

However, when $w(x, y)$ has the form of (3.39) this then implies that:

$$\phi_n(0) = \phi_n''(0) = 0 \tag{3.50}$$

Similarly, appropriate expressions are found for the clamped edge and the free edge. Then, whatever the relevant form of the boundary conditions on $x = 0$ and $x = a$, the total $\phi_n(x) = \phi_{n_H} + \phi_{n_p}$ and hence $w(x, y)$ is known from (3.39). Then, for a composite material laminated plate, one must calculate the curvatures

$$\kappa_x = -\frac{\partial^2 w}{\partial x^2}, \quad \kappa_y = -\frac{\partial^2 w}{\partial y^2}, \quad \text{and} \quad \kappa_{xy} = -\frac{\partial^2 w}{\partial x \partial y}$$

Knowing these, one can calculate the stresses in each of the k laminae through the following:

$$\begin{bmatrix} \sigma_x \\ \sigma_y \\ \sigma_{xy} \end{bmatrix}_k = \begin{bmatrix} \overline{Q}_{11} & \overline{Q}_{12} & 0 \\ \overline{Q}_{12} & \overline{Q}_{22} & 0 \\ 0 & 0 & 2\overline{Q}_{66} \end{bmatrix}_k \begin{bmatrix} \kappa_x \\ \kappa_y \\ \kappa_{xy} \end{bmatrix} z \tag{3.51}$$

The stresses thus derived for each lamina must then be compared with the allowable stresses, determined through some failure criterion (discussed in Chapter 7) to see if structural integrity is retained under a given load (the analysis problem) or if this set of materials and orientation is sufficient for an objective load (the design problem).

Finally, the Levy type solution is fine for a composite plate with no bending stretching coupling, with mid-plane symmetry and with two opposite edges simply supported. If two opposite edges are not simply supported then the complexity of the functions necessary to satisfy the y boundary condition causes problems. Such functions will be discussed in Chapter 6.

3.8. Perturbation Solutions for the Bending of a Composite Material Plate, with Mid-Plane Symmetry and No Bending Twisting Coupling

As shown in the previous two sections, the Navier approach is excellent for plates with all four edges simply supported, and the Levy approach is fine for plates with two opposite edges simply supported, regardless of the boundary conditions on the other two edges. But for a composite plate with two opposite edges simply supported, even the Levy approach yields three distinct solutions depending on the relative magnitudes of D_1, D_2 and $D_3 = D_{12} + 2D_{66}$. In addition, there are numerous books and papers, available for the solution of isotropic plate problems, a list of which is readily available [1,2].

Aware of all of the above, and based upon the fact that the solution of the second case of the Levy solution of (3.47) has the same form as that of the isotropic case, Vinson showed that the cases of (3.46) and (3.48) can be dealt with as perturbations about the solution of the same plates composed of isotropic materials [4,5].

Consider the governing equation for the bending of a composite material exhibiting mid-plane symmetry ($B_{ij} = 0$), no bending twisting coupling ($D_{16} = D_{26} = 0$), and no transverse shear deformation (that is (3.23) applies). Then equation, (3.29) is after dividing both sides by D_1:

$$\frac{\partial^4 w}{\partial x^4} + \frac{2D_3}{D_1} \frac{\partial^4 w}{\partial x^2 \partial y^2} + \frac{D_2}{D_1} \frac{\partial^4 w}{\partial y^4} = \frac{p(x, y)}{D_1} \tag{3.52}$$

If one uses coordinate stretching such that the following is defined; let

$$\bar{y} = \left(\frac{D_2}{D_1}\right)^{-1/4} y \quad , \qquad \bar{b} = \left(\frac{D_2}{D_1}\right)^{-1/4} b \tag{3.53}$$

substituting (3,53) into (3.52) yields

$$\frac{\partial^4 w}{\partial x^4} + 2\left(\frac{D_2}{D_1}\right)^{-1/2} \cdot \left(\frac{D_3}{D_1}\right)\frac{\partial^4 w}{\partial x^2 \partial \bar{y}^2} + \frac{\partial^4 w}{\partial \bar{y}^4} = \frac{p(x,\bar{y})}{D_1}. \tag{3.54}$$

Next defining a quantity α to be

$$\alpha = 2\left[1 - \left(\frac{D_2}{D_1}\right)^{-1/2} \cdot \left(\frac{D_3}{D_1}\right)\right] \tag{3.55}$$

it is seen that substituting (3.55) in (3.54) yields:

$$\frac{\partial^4 w}{\partial x^4} + \frac{2\partial^4 w}{\partial x^2 \partial \bar{y}^2} + \frac{\partial^4 w}{\partial \bar{y}^4} - \frac{\alpha \partial^4 w}{\partial x^2 \partial \bar{y}^2} = \frac{p(x,\bar{y})}{D_1} \tag{3.56}$$

If one defines the biharmonic operator, used in all isotropic plate problems, to be, in the stretched coordinate system:

$$\nabla^4 w = \frac{\partial^4 w}{\partial x^4} + \frac{2\partial^4 w}{\partial x^2 \partial \bar{y}^2} + \frac{\partial^4 w}{\partial \bar{y}^4} \tag{3.57}$$

then (3.56) becomes

$$\nabla^4 w - \frac{\alpha \partial^4 w}{\partial x^2 \partial \bar{y}^2} = \frac{p(x,y)}{D_1} \tag{3.58}$$

Finally, assume the form of the solution for $w(x,\bar{y})$ to be

$$w(x,\bar{y}) = \sum_{n=0}^{\infty} w_n(x,\bar{y})\alpha^n \tag{3.59}$$

which is a perturbation solution employing the "small" parameter α. Substituting (3.59) into (3.58) and equating all coefficients of α^n to zero, it is easily found that:

$$\nabla^4 w_0 = \frac{p(x,\bar{y})}{D_1} \tag{3.60}$$

$$\nabla^4 w_n = -\frac{\partial^4 w_{n-1}(x,\bar{y})}{\partial x^2 \partial \bar{y}^2} \qquad n \geq 1 \tag{3.61}$$

It is seen that (3.60) is the governing differential equation for an isotropic plate of stiffness D_1, subjected to the actual lateral load

$p(x, \bar{y})$, with the stretched coordinate $\bar{y}$ defined in (3.53). It is probable that, regardless of boundary conditions on any edge, the solution is available, either exactly or approximately. Then all subsequent w's, such as w_1, w_2, w_3, and so on are available from solving (3.61) a plate whose lateral load is $-D_1(\partial^4 w_{n-1}/\partial x^2 \partial \bar{y}^2)$, where w_{n-1} is known, whose flexural stiffness is D_1 and whose boundary conditions are the same as those of the actual plate.

This technique is very useful because the "small" perturbation parameter need not be so small; it has been proven that when $|\alpha| < 1$, (3.59) is another form of the exact solution, and $|\alpha| < 1$ covers much of the practical composite material construction regime. Also from a computational point of view, it is seldom necessary to include terms past $n = 1$, namely from (3.59):

$$w(x, \bar{y}) = w_0 + w_1 \alpha \tag{3.62}$$

Not only is the above technique useful, but if $|\alpha| \geqslant 1$, then the composite may fall within another range where for $(D_2/D_1) \ll 1$, the plate behaves as a plate in the x-direction, but because $(D_2/D_1) \ll 1$, it behaves as a membrane in the y-direction, with the following simpler governing differential equation,

$$\frac{\partial^4 w}{\partial x^4} + 2 \frac{D_3}{D_1} \frac{\partial^4 w}{\partial x^2 \partial y^2} = \frac{p(x, y)}{D_1} \tag{3.63}$$

with the solution in the form of the following for a plate simply supported on the y edges

$$w(x, y) = \sum_{n=1}^{\infty} \phi_n(x) \sin \lambda_n^* y^* \tag{3.64}$$

where

$$y^* = \left(\frac{2D_3}{D_1}\right)_y^{-1/2}, \quad \lambda_n^* = \frac{n\pi}{b^*}, \quad b^* = \left(\frac{2D_3}{D_1}\right)^{-1/2} b$$

Finally, if $(\bar{b}/a) = (D_2/D_1)^{-1/4}(b/a) > 3$, then the plate behaves purely as a beam in the x-direction as discussed later in Chapter 4, regardless of the boundary conditions on the y edge, as far as maximum deflection and maximum stresses. Hence, beam solutions have easy application to the solution of many composite plate problems. All of the details of the last two techniques are given in detail in references 4 and 5.

Incidentally, the techniques described in this section are the first use of perturbation techniques involving material property perturbation, even though geometric perturbations have been utilized for many decades.

3.9. Perturbation Solutions for the Bending of a Composite Material Plate with Bending-Twisting Coupling and Mid-Plane Symmetry

In 1971, Farshad and Ahmadi [4] expanded the use of perturbation solutions through studying composite material plates which have bending twisting coupling, that is, $D_{16} \neq 0$, $D_{26} \neq 0$, by perturbing the solutions of plates without such coupling such as those of the previous section, and/or those which are described by (3.29).

Looking at (2.62) the moment curvature relations for a rectangular plate with bending twisting coupling are:

$$M_x = D_{11}\kappa_x + D_{12}\kappa_y + 2D_{16}\kappa_{xy}$$

$$M_y = D_{12}\kappa_x + D_{22}\kappa_y + 2D_{26}\kappa_{xy}$$

$$M_{xy} = D_{16}\kappa_x + D_{26}\kappa_y + 2D_{66}\kappa_{xy}$$

where as before, if transverse shear deformation is ignored

$$\kappa_x = -\frac{\partial^2 w}{\partial x^2}, \quad \kappa_y = -\frac{\partial^2 w}{\partial y^2} \quad \text{and} \quad \kappa_{xy} = -\frac{\partial^2 w}{\partial x \partial y}$$

Therefore the moment curvature relations become:

$$M_x = -D_{11}\frac{\partial^2 w}{\partial x^2} - D_{12}\frac{\partial^2 w}{\partial y^2} - 2D_{16}\frac{\partial^2 w}{\partial x \partial y} \tag{3.65}$$

$$M_y = -D_{12}\frac{\partial^2 w}{\partial x^2} - D_{22}\frac{\partial^2 w}{\partial y^2} - 2D_{26}\frac{\partial^2 w}{\partial x \partial y} \tag{3.66}$$

$$M_{xy} = -D_{16}\frac{\partial^2 w}{\partial x^2} - D_{26}\frac{\partial^2 w}{\partial y^2} - 2D_{66}\frac{\partial^2 w}{\partial x \partial y} \tag{3.67}$$

Compare these with (3.24) through (3.26). Substituting the above into the equilibrium equation (3.19), which holds for any rectangular plate, gives

$$D_{11}\frac{\partial^4 w}{\partial x^4} + D_{16}\frac{\partial^4 w}{\partial x^3 \partial y^1} + 2(D_{12} + 2D_{66})\frac{\partial^4 w}{\partial x^2 \partial y^2}$$

$$+ 4D_{26}\frac{\partial^4 w}{\partial x \partial y^3} + D_{22}\frac{\partial^4 w}{\partial y^4} = p(x, y) \tag{3.68}$$

Comparing this with (3.27), it is seen that due to the existence of D_{16} and D_{26}, odd numbered derivatives appear in the governing differential equation. That, of course, precludes the use of both the Navier approach of Section 3.5 and the use of the Levy approach of Section 3.7.

Farshad and Ahmadi restrict themselves to a plate of one lamina, hence from Chapter 2, $h_0 = -h/2$, $h_1 = +h/2$, where h is the plate thickness (see Figure 3.5). Therefore

$$D_{ij} = \frac{\overline{Q}_{ij} h^3}{12}$$

and from the expressions of Chapter 2 one can be assured of the following, where $m = \cos \theta$, and $n = \sin \theta$ (see Figure 2.8)

$$D_{11} = D_1 m^4 + D_2 n^4 + 2 D_3 m^2 n^2$$

$$D_{22} = D_1 n^4 + D_2 m^4 + 2 D_3 m^2 n^2$$

$$D_{16} = mn \left\{ - D_2 n^2 + D_1 m^2 - D_3 (m^2 - n^2) \right\} \qquad (3.69)$$

$$D_{26} = mn \left\{ - D_2 m^2 + D_1 n^2 + D_3 (m^2 - n^2) \right\}$$

$$D_{12} = (D_1 + D_2 - 2 D_{66}) m^2 n^2 + D_{12} (m^4 + n^4)$$

$$D_{66} = (D_1 + D_2 - 2 D_{12}) m^2 n^2 + D_{66} (m^2 - n^2)^2$$

which agree exactly with (2.62) if for the single layer plate one defines the following which is consistent with Section 2.6:

$$D_1 = \frac{E_1 h^3}{12(1 - \nu_{12}\nu_{21})}, \quad D_2 = \frac{E_2 h^3}{12(1 - \nu_{12}\nu_{21})}, \quad D_3 = D_{12} + 2 D_{66}$$

$$D_{12} = D_1 \nu_{21} = D_2 \nu_{12} \quad D_{66} = \frac{G_{12} h^3}{12}$$

Suppose θ is sufficiently small that we can expand $\sin \theta$ and $\cos \theta$ in power series of θ, and retain the first few terms as an approximation. In that case,

$$D_{11} = D_1 + 2\alpha_1 \theta^2 + \beta_1 \theta^4 + \dots$$
$$D_{22} = D_2 + 2\alpha_2 \theta^2 + \beta_2 \theta^4 + \dots$$
$$D_{16} = - \alpha_1 \theta - \beta_1 \theta^3 + \gamma_1 \theta^5 + \dots$$
$$D_{26} = \alpha_2 \theta + \beta_2 \theta^3 - \gamma_2 \theta^5 + \dots \qquad (3.70)$$
$$D_{12} = D_{12} + \alpha_3 \theta^2 + \beta_3 \theta^4 + \dots$$
$$D_{66} = D_{66} + \alpha_3 \theta^2 + \beta_3 \theta^4 + \dots$$

in which

$$\alpha_1 = D_3 - D_1 \qquad \beta_1 = -\left(\tfrac{5}{3}\alpha_1 + d_2\right) \qquad \gamma_1 = -\left(\alpha_1 + \alpha_2 + \tfrac{2}{15}\alpha_1\right)$$

$$\alpha_2 = D_3 - D_2 \qquad \beta_2 = -\left(\alpha_1 + \tfrac{5}{3}\alpha_2\right) \qquad \gamma_2 = -\left(\alpha_1 + \alpha_2 + \tfrac{2}{15}\alpha_2\right)$$

$$\alpha_3 = -(\alpha_1 + \alpha_2) \qquad \beta_3 = \tfrac{4}{3}(\alpha_1 + \alpha_2)$$

Substituting (3.70) into (3.68) and retaining terms up through θ^4, we obtain

$$L(w) + \theta M(w) + \theta^2 N(w) + \theta^3 P(w) + \theta^4 Q(w) = p(x, y) \qquad (3.71)$$

where the differential operators are given by

$$L(w) = D_1 \frac{\partial^4 w}{\partial x^4} + 2 D_3 \frac{\partial^4 w}{\partial x^2 \partial y^2} + D_2 \frac{\partial^4 w}{\partial y^4} \qquad (3.72)$$

$$M(w) = 4\alpha_1 \frac{\partial^4 w}{\partial x^3 \partial y} - 4\alpha_2 \frac{\partial^4 w}{\partial x \partial y_3} \qquad (3.73)$$

$$N(w) = 2\alpha_1 \frac{\partial^4 w}{\partial x^4} + 6\alpha_3 \frac{\partial^4 w}{\partial x^2 \partial y^2} + 2\alpha_2 \frac{\partial^4 w}{\partial y^4} \qquad (3.74)$$

$$P(w) = 4\beta_1 \frac{\partial^4 w}{\partial x^3 \partial y} - 4\beta_2 \frac{\partial^4 w}{\partial x \partial y^3} \qquad (3.75)$$

$$Q(w) = \beta_1 \frac{\partial^4 w}{\partial x^4} + 6\beta_3 \frac{\partial^4 w}{\partial x^2 \partial y^2} + \beta_2 \frac{\partial^4 w}{\partial y^4} \qquad (3.76)$$

Obviously retaining terms through θ^4 is arbitrary, and depending on the size of θ, one could retain either more or fewer powers of θ.

Since θ is assumed small, we can assume a perturbation solution to (3.71) of the following form.

$$w = w_0 + w_1\theta + w_2\theta^2 + w_3\theta^3 + w_4\theta^4 = \sum_{n=0}^{N} w_n\theta^n \qquad (3.77)$$

Substituting (3.77) into (3.71) and equating coefficients of like powers of θ to zero, one obtains

$$L(w_0) = p(x, y) \qquad (3.78)$$

$$L(w_1) = -M(w_0) \qquad (3.79)$$

$$L(w_2) = -M(w_1) - N(w_0) \tag{3.80}$$

$$L(w_3) = -M(w_2) - N(w_1) - P(w_0) \tag{3.81}$$

$$L(w_4) = -M(w_3) - N(w_2) - P(w_1) - Q(w_0) \tag{3.82}$$

Hence, in each case w_n is found by sequentially solving plate equations of the form given in (3.72) which of course is identical to solving (3.29), hence, the Navier approach, the Levy approach, or the approach of Section 3.8 can be used depending upon the boundary conditions. In the first of the above, one has as the forcing function the actual loading, $p(x, y)$. In the subsequent solutions one finds w_n using the same differential equation, but the forcing functions involve derivatives of previously found w_m and $m < n$.

As in the perturbation solution of the preceding section, any non-homogeneous boundary condition should be satisfied in the w_0 solution, so that for $n = 1, 2, \ldots$, all boundary conditions are homogeneous.

3.10. Eigenvalue Problems of Plates of Composite Materials: Natural Vibrations and Elastic Stability

To this point the problems studied in this Chapter have concentrated on finding the maximum deflection in composite material beams and plates to insure that they are not too large for a deflection limited or stiffness critical structure, and in determining the maximum stresses in the beam or plate structure for those which are strength critical. However, there are two other ways in which a structure can become damaged or useless; one is through a dynamic response to time dependent loads, resulting again to too large a deflection or too high stresses, and the other is through the occurrence of an elastic instability (buckling).

In the former, dynamic loading on a structure can vary from a reoccurring cycling loading of the same repeated magnitude, such as a structure supporting an unbalanced motor that is turning at 100 revolutions per minute, for example, to the other extreme of a short time intense, non-reoccurring load, termed shock or impact loading, such as a bird striking an aircraft component during flight. A continuous infinity of dynamic loads exist between these extremes of harmonic oscillation and impact.

A whole volume could and should be written on the dynamic response of composite material structures to time dependent loads, and that is beyond the scope of this text. There are a number of texts dealing with dynamic response of isotropic structures. However, one common thread to all dynamic response will be presented - that of the natural frequencies of vibration of composite beams and plates and shells.

Any continuous structure mathematically has an infinity of natural frequencies and mode shapes. If a structure is oscillated at a frequency that corresponds to a natural frequency, it will respond by a rapidly growing amplitude with time, requiring very little input energy, until such time as the structure becomes overstressed and fails, or until the oscillations become so large that non-linear effects may limit the amplitude to a large but usually unsatisfactory one because then considerable fatigue damage will occur.

Thus, for any structure, in analyzing it for structural integrity, the natural frequencies should be determined in order to compare them with any time dependent loadings to which the structure will be subjected to insure that the frequencies imposed and the natural frequencies differ considerably. Conversely in designing a structure, over and above insuring that the structure will not over-deflect, or become overstressed, care should be taken to avoid resonances (that is, imposed loads having the same frequency as one or more natural frequencies).

The easiest example to illustrate this behavior is that of the composite material plate previously studied, wherein midplane symmetry exists ($B_{ij} = 0$), no other coupling terms, $(\ )_{16} = (\ )_{26} = 0$, and no transverse shear deformation exist ($\epsilon_{xz} = \epsilon_{yz} = 0$). In this case the governing equations are given by (3.29) as shown below.

$$D_1 \frac{\partial^4 w}{\partial x^4} + 2D_3 \frac{\partial^4 w}{\partial x^2 \partial y^2} + D_2 \frac{\partial^4 w}{\partial y^4} = p(x, y). \tag{3.83}$$

If D'Alambert's Principle were used then one could add a term to (3.83) equal to the product of mass per unit area and acceleration per unit length. In that case, the right-hand side of (3.83) becomes

$$p(x, y, t) - \frac{\rho h \partial^2 w(x, y, t)}{\partial t^2} \tag{3.84}$$

where here w and p both become functions of time as well as space, ρ is the *mass* density of the material, and h is the plate thickness. In the above $p(x, y, t)$ now is the spatially varying time dependent forcing function causing the dynamic response, and could be anything from a harmonic oscillation to an intense one time impact.

If the plate is a hybrid (differing materials across the thickness) composite, it is the average of the mass density across the thickness, that is

$$\rho = \frac{1}{h} \sum_{k=1}^{N} \rho_k (h_k - h_{k-1}), \tag{3.85}$$

that must be used in (3.84).

To investigate *natural* vibration behavior, the forcing function $p(x, y, t)$ is taken to be zero.

Again, we will assume that the vibrational mode shapes of a composite material plate simply supported on all four sides is identical to an isotropic plate with the same boundary conditions. A time dependent function is included below to represent the harmonic oscillation expected in linear vibration.

$$w(x, y, t) = \sum_{m=1}^{\infty} \sum_{n=1}^{\infty} A_{mn} \sin\frac{m\pi x}{a} \sin\frac{n\pi y}{b} \cos \omega_{nt}. \tag{3.86}$$

Substituting these mode shapes into (3.83) and (3.84) with $p(x, y, t) = 0$ provides the corresponding natural frequencies:

$$\omega_{mn} = \frac{\pi^2}{\sqrt{\rho h}} \sqrt{D_1\left(\frac{m}{a}\right)^4 + 2D_3\left(\frac{m}{a}\right)^2\left(\frac{n}{b}\right)^2 + D_2\left(\frac{n}{b}\right)^4} \tag{3.87}$$

The fundamental frequency occurs with $m = n = 1$, which is for one half sine wave in each direction. Also note that the amplitude A_{mn} cannot be determined from this eigenvalue theory, and that (3.83) and (3.84) form a homogeneous equation when $p(x, y, t) = 0$.

In addition to looking at maximum deflections, maximum stresses and natural frequencies when analyzing a structure, one must investigate under what loading conditions an instability can occur, which is also generically referred to as buckling.

For a plate there are five equations associated with the in-plane loads N_x, N_y and N_{xy} and the in-plane displacements they cause, u_0 and v_0. These equations are given by (3.9), (3.10), and from (2.62) for the case of midplane symmetry ($B_{ij} = 0$) it is seen that

$$N_x = A_{11}\epsilon_x^0 + A_{12}\epsilon_y^0 + 2A_{16}\epsilon_{xy}^0 \tag{3.88}$$

$$N_y = A_{12}\epsilon_x^0 + A_{22}\epsilon_y^0 + 2A_{26}\epsilon_{xy}^0 \tag{3.89}$$

$$N_{xy} = A_{16}\epsilon_x^0 + A_{26}\epsilon_y^0 + 2A_{64}\epsilon_{xy}^0 \tag{3.90}$$

Likewise the six governing equations involving M_x, M_y, M_{xy}, Q_x, Q_y and w, are given by (3.12), (3.14), (3.15), (3.65), (3.66) and (3.67). At any rate there is no coupling between in-plane and lateral action for the plate with mid-plane symmetry. Yet it is well known and often observed that in-plane loads through buckling do cause lateral deflections, which are usually disastrous.

The answer to the paradox is that entirely through this Chapter up to this point we have used a *linear* elasticity theory, and the physical event of buckling is a non-linear problem. For brevity, the development of the non-linear theory will not be included herein because it is included in so many other texts including those already referenced earlier [1,2].

The results of including the terms to predict the advent or inception of buckling for the plate are, modifying (3.83)

$$D_1 \frac{\partial^4 w}{\partial x^4} + 2D_3 \frac{\partial^4 w}{\partial x^2 \partial y^2} + D_2 \frac{\partial^4 w}{\partial y^4} = p(x, y) + N_x \frac{\partial^2 w}{\partial x^2}$$

$$+ 2N_{xy} \frac{\partial^2 w}{\partial x \partial y} + N_y \frac{\partial^2 w}{\partial y^2} \qquad (3.91)$$

where clearly there is a coupling between the in-plane loads and the lateral deflection.

The buckling loads like the natural frequencies are independent of the lateral loads, which will be disregarded in what follows. However, in actual structural analysis, the effect of lateral loads, along with the in-plane loads could cause overstressing and failure before the in-plane buckling load is reached. However, the buckling load is independent of the lateral load, as are the natural frequencies. Incidentally, common sense dictates that if one is designing a structure to withstand compressive loads, with the possibility of buckling being the failure mode, one had better design the structure to be midplane symmetric, so that $B_{ij} = 0$, otherwise the bending-stretching coupling would likely cause overstressing before the buckling load is reached.

Looking now at (3.91) for the buckling of the composite plate under an axial load N_x only, and ignoring $p(x, y)$ the equation becomes:

$$D_1 \frac{\partial^4 w}{\partial x^4} + 2D_3 \frac{\partial^4 w}{\partial x^2 \partial y^2} + D_2 \frac{\partial^4 w}{\partial y^4} - N_x \frac{\partial^2 w}{\partial x^2} = 0 \qquad (3.92)$$

Again assuming the buckling mode for a laminated composite material plate to be that of an isotropic plate with the same boundary conditions, then for the case of the plate simply supported on all four edges one assumes

$$w(x, y) = \sum_{m=1}^{\infty} \sum_{n=1}^{\infty} A_{mn} \sin \frac{m\pi \gamma}{a} \sin \frac{n\pi y}{b} \qquad (3.93)$$

Substituting (3.93) into (3.92) it is seen that for the critical load $N_{x_{cr}}$

$$N_{x_{cr}} = -\frac{\pi^2 a^2}{m^2} \left[D_1 \left(\frac{m}{a} \right)^4 + 2 D_3 \left(\frac{m}{a} \right)^2 \left(\frac{n}{b} \right)^2 + D_2 \left(\frac{n}{b} \right)^4 \right] \qquad (3.94)$$

where again several things are clear: (3.92) is a homogeneous equation, one cannot determine the value of A_{mn}, and again only the lowest value of $N_{x_{cr}}$ is of any importance usually. However it is not clear which value of m and n result in the lowest critical buckling load. All values of n appear in the numerator, so $n = 1$ is the necessary value. But m appears several places, and depending upon the value of the flexural stiffness D_1, D_2 and D_3, and the length to width ratio of the plate, a/b, it is not clear which value of m will provide the lowest value of $N_{x_{cr}}$. However for a given plate it is easily determined computationally.

Thus, it is seen that the eigenvalue problems differ from the considerations in previous sections because the event, natural vibration or buckling, occur at only certain values. Hence, the natural frequencies and the buckling load are the eigenvalues; the vibrational mode and the buckling mode are the eigenfunctions.

In analyzing any structure one therefore should determine four things: the maximum deflection, the maximum stresses, the natural frequencies (if there is any dynamic loading to the structure - or nearby to the structure) and the buckling loads (if there are any compressive loads).

What about the natural vibrations and buckling loads of composite material plates with boundary conditions other than simply supported? It is seen that for the cases treated in this section that double sine series were used because those are the vibrational modes and buckling modes of plates simply supported on all edges.

All combinations of beam vibrational mode shapes are applicable for use for plates with various boundary conditions. These have been developed by Warburton [7], and all derivatives and integrals of those functions catalogued conveniently by Young and Felgar [8,9] for easy use.

Similar expressions for buckling and vibrational modes are in general available also in many texts, so likewise they can be used for (3.86) and (3.93).

It must again be stated that the natural frequencies and buckling loads calculated in this section do not include transverse shear effects, and are therefore only approximate - but they are useful for preliminary design, because of their relative simplicity. If the transverse shear deformation were included, see the next Section, both natural frequencies and buckling loads would be lower than those calculated in this Section - so the buckling loads calculated, neglecting transverse shear deformation, are not conservative.

3.11. Static and Dynamic Analysis of Plates of Composite Materials
Including Transverse Shear Deformation Effects

In the preceding Sections of Chapter 3, all of the analyses have neglected transverse shear deformation effects, which means that those analyses and results are approximate, but because of their relative simplicity are useful for preliminary design for "sizing" the structure. Now these effects will be taken into account for a composite material plate subjected to static loads and dynamic loads. The neglect of transverse shear deformation simply means that, as stated previously

$$\epsilon_{xz} = \frac{1}{2}\left(\frac{\partial u}{\partial z} + \frac{\partial w}{\partial x}\right) = 0$$
$$\epsilon_{yz} = \frac{1}{2}\left(\frac{\partial v}{\partial z} + \frac{\partial w}{\partial y}\right) = 0$$

$$(3.95)$$

Since the in-plane displacements $u(x, y, z)$ and $v(x, y, z)$ have at most a translation and a rotation, their most general form is

$$u(x, y, z) = u_0(x, y) + z\bar{\alpha}(x, y)$$
$$v(x, y, z) = v_0(x, y) + z\bar{\beta}(x, y)$$

$$(3.96)$$

then from (3.95) it is seen that

$$\bar{\alpha} = -\frac{\partial w}{\partial x} \quad \text{and} \quad \bar{\beta} = -\frac{\partial w}{\partial y}$$

$$(3.97)$$

and since

$$\kappa_x = \frac{\partial \bar{\alpha}}{\partial x}, \quad \kappa_y = \frac{\partial \bar{\beta}}{\partial y} \quad \text{and} \quad \kappa_{xy} = \frac{1}{2}\left(\frac{\partial \bar{\alpha}}{\partial y} + \frac{\partial \bar{\beta}}{\partial x}\right)$$

$$(3.98)$$

then

$$\kappa_x = -\frac{\partial^2 w}{\partial x^2}, \quad \kappa_y = -\frac{\partial^2 w}{\partial y^2} \quad \text{and} \quad \kappa_{xy} = -\frac{\partial^2 w}{\partial x \partial y}$$

$$(3.99)$$

as was shown previously in (3.23).

Looking now at the plate equilibrium equations and constitutive equations, the complete set is given as follows, for the static case:

$$\frac{\partial N_x}{\partial x} + \frac{\partial N_{xy}}{\partial y} + \tau_{1x} - \tau_{2x} = 0$$

$$(3.100)$$

$$\frac{\partial N_{xy}}{\partial x} + \frac{\partial N_y}{\partial y} + \tau_{1y} - \tau_{2y} = 0 \tag{3.101}$$

$$\frac{\partial Q_x}{\partial x} + \frac{\partial Q_y}{\partial y} + p(x, y) = 0 \tag{3.102}$$

$$\frac{\partial M_x}{\partial x} + \frac{\partial M_{xy}}{\partial y} - Q_x + \frac{h}{2}[\tau_{1x} + \tau_{2x}] = 0 \tag{3.103}$$

$$\frac{\partial M_{xy}}{\partial x} + \frac{\partial M_y}{\partial y} - Q_y + \frac{h}{2}[\tau_{1y} + \tau_{2y}] = 0 \tag{3.104}$$

For a plate that is mid-plane symmetric ($B_{ij} = 0$), and has no coupling [($_{16}$) = ($_{24}$) = ($_{45}$) = 0], the constitutive equations are

$$N_x = A_{11}\epsilon_x^0 + A_{12}\epsilon_y^0 \tag{3.105}$$

$$N_y = A_{12}\epsilon_x^0 + A_{22}\epsilon_y^0 \tag{3.106}$$

$$N_{xy} = 2A_{66}\epsilon_{xy}^0 \tag{3.107}$$

$$M_x = D_{11}\kappa_x + D_{12}\kappa_y \tag{3.108}$$

$$M_y = D_{12}\kappa_x + D_{22}\kappa_y \tag{3.109}$$

$$M_{xy} = 2D_{66}\kappa_{xy} \tag{3.110}$$

$$Q_x = 2A_{55}\epsilon_{xz} = 2A_{55}\left(\bar{\alpha} + \frac{\partial w}{\partial x}\right) \tag{3.111}$$

$$Q_y = 2A_{44}\epsilon_{yz} = 2A_{44}\left(\bar{\beta} + \frac{\partial w}{\partial y}\right) \tag{3.112}$$

Because the plate is mid-plane symmetric there is no bending-stretching coupling, hence the in-plane loads and deflections are uncoupled (separate) from the lateral loads, deflection and rotation. Hence, for the lateral action, (3.102) through (3.104) and (3.108) through (3.112) are utilized: 8 equations and 8 unknowns.

Substituting (3.108) through (3.112) into (3.102) through (3.104) and using (3.98) results in the following set of governing differential equations for a laminated composite plate subjected to a lateral load, wherein ($B_{ij} = 0$), ( )$_{16}$ = ( )$_{26}$ = ( )$_{45}$ = 0, and no surface shear stresses.

$$D_{11}\frac{\partial^2 \bar{\alpha}}{\partial x^2} + D_{66}\frac{\partial^2 \bar{\alpha}}{\partial y^2} + (D_{12} + D_{66})\frac{\partial^2 \bar{\beta}}{\partial x\partial y} - 2A_{55}\left(\bar{\alpha} + \frac{\partial w}{\partial x}\right) = 0 \tag{3.113}$$

$$(D_{12} + D_{66})\frac{\partial^2\bar{\alpha}}{\partial x \partial y} + D_{66}\frac{\partial^2\bar{\beta}}{\partial x^2} + D_{22}\frac{\partial^2\bar{\beta}}{\partial y^2} - 2A_{44}\left(\bar{\beta} + \frac{\partial w}{\partial y}\right) = 0 \qquad (3.114)$$

$$2A_{55}\left(\frac{\partial\bar{\alpha}}{\partial x} + \frac{\partial^2 w}{\partial x^2}\right) + 2A_{44}\left(\frac{\partial\bar{\beta}}{\partial y} + \frac{\partial^2 w}{\partial y^2}\right) + p(x, y) = 0 \qquad (3.115)$$

Dobyns [10] has employed the Navier approach to solving these equations for a composite plate simply supported on all four edges subjected to a lateral load, using the following functions

$$w(x, y) = \sum_{m=1}^{\infty}\sum_{n=1}^{\infty} C_{mn} \sin\frac{m\pi x}{a} \sin\frac{n\pi y}{b} \qquad (3.116)$$

$$\bar{\alpha}(x, y) = \sum_{m=1}^{\infty}\sum_{n=1}^{\infty} A_{mn} \cos\frac{m\pi x}{a} \sin\frac{n\pi y}{b} \qquad (3.117)$$

$$\bar{\beta}(x, y) = \sum_{m=1}^{\infty}\sum_{n=1}^{\infty} B_{mn} \sin\frac{m\pi x}{a} \cos\frac{n\pi y}{b} \qquad (3.118)$$

$$p(x, y) = \sum_{m=1}^{\infty}\sum_{n=1}^{\infty} q_{mn} \sin\frac{m\pi x}{a} \sin\frac{n\pi y}{b} \qquad (3.119)$$

which satisfy the boundary conditions on all edges such that

$$w = 0, \quad \frac{\partial\bar{\alpha}}{\partial x} = 0 \quad \text{on} \quad x = 0, a \quad \text{and} \quad \frac{\partial\bar{\beta}}{\partial y} = 0 \quad \text{on} \quad y = 0, b$$

Substituting these functions into the governing differential equations above result in the following:

$$\begin{bmatrix} L_{11} & L_{12} & L_{13} \\ L_{12} & L_{22} & L_{23} \\ L_{13} & L_{23} & L_{33} \end{bmatrix}\begin{Bmatrix} A_{mn} \\ B_{mn} \\ C_{mn} \end{Bmatrix} = \begin{Bmatrix} 0 \\ 0 \\ q_{mn} \end{Bmatrix} \qquad (3.120)$$

where, if $\lambda_m = m\pi/a$ and $\lambda_n = n\pi/b$, and q_{mn} is the lateral load coefficient

$$L_{11} = D_{11}\lambda_m^2 + D_{66}\lambda_n^2 + 2A_{55}$$

$$L_{12} = (D_{12} + D_{66})\lambda_m\lambda_n$$

$$L_{13} = 2A_{55}\lambda_m$$

$$L_{22} = D_{66}\lambda_m^2 + D_{22}\lambda_n^2 + 2A_{44}$$

$$L_{23} = 2A_{44}\lambda_n$$

$$L_{33} = 2A_{55}\lambda_m^2 + 2A_{44}\lambda_n^2$$

Solving (3.120), one obtains

$$A_{mn} = \frac{(L_{12}L_{23} - L_{22}L_{13})q_{mn}}{\det} \tag{3.121}$$

$$B_{mn} = \frac{(L_{12}L_{13} - L_{11}L_{23})q_{mn}}{\det} \tag{3.122}$$

$$C_{mn} = \frac{(L_{11}L_{22} - L_{12}^2)q_{mn}}{\det} \tag{3.123}$$

where det is the determinant of the matrix in Equation (3.120).

Having solved the problem to obtain $\bar{\alpha}$, $\bar{\beta}$ and w, the curvatures $\kappa_x = (\partial\bar{\alpha}/\partial x)$, $\kappa_y = (\partial\bar{\beta}/\partial y)$ and $\kappa_{xy} = 1/2[(\partial\bar{\alpha}/\partial y) + (\partial\bar{\beta}/\partial y)]$ may be obtained. Then for a laminated composite plate the stresses in each lamina may be obtained by:

$$\begin{bmatrix} \sigma_x \\ \sigma_y \\ \sigma_{xy} \end{bmatrix}_k = \begin{bmatrix} \bar{Q}_{11} & \bar{Q}_{12} & 0 \\ \bar{Q}_{12} & \bar{Q}_{22} & 0 \\ 0 & 0 & 2\bar{Q}_{66} \end{bmatrix}_k z \begin{bmatrix} \kappa_x \\ \kappa_y \\ \kappa_{xy} \end{bmatrix} \tag{3.124}$$

If the lateral load $p(x, y)$ is distributed over the entire lateral surface, then

$$q_{mn} = \frac{4}{ab} \int_0^a \int_0^b p(x, y) \sin\frac{m\pi x}{a} \sin\frac{n\pi y}{b} dx dy \tag{3.125}$$

If that load is uniform then

$$q_{mn} = \frac{4p(x, y)}{mn\pi^2}(1 - \cos m\pi)(1 - \cos n\pi) \tag{3.126}$$

For a concentrated load located at $x = \xi$ and $y = \eta$,

$$q_{mn} = \frac{4P}{ab} \sin\frac{m\pi\xi}{a} \sin\frac{n\pi\eta}{b} \tag{3.127}$$

where P is the total load.

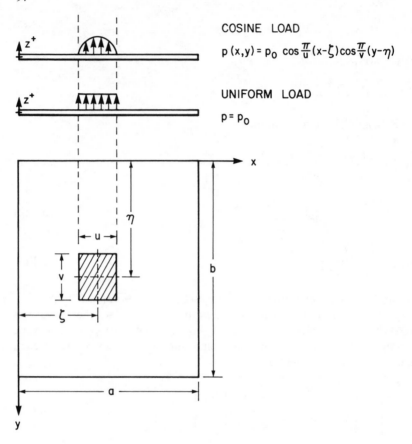

COSINE LOAD

$p(x,y) = p_0 \cos\frac{\pi}{u}(x-\zeta)\cos\frac{\pi}{v}(y-\eta)$

UNIFORM LOAD

$p = p_0$

Figure 3.8.

For loads over a rectangular area of side lengths u and v whose center is at ξ and η, as shown in Figure 3.8 q_{mn} is given as follows:

For $\dfrac{n}{b} \neq \dfrac{1}{v}$ and $\dfrac{m}{a} \neq \dfrac{1}{u}$, then

$$q_{mn} = \frac{4p(xy)\sin\left(\dfrac{m\pi\eta}{b}\right)\cos\left(\dfrac{n\pi v}{2b}\right)\cos\left(\dfrac{m\pi u}{2a}\right)}{abu^2v^2\left(\dfrac{n}{b}-\dfrac{1}{v}\right)\left(\dfrac{n}{b}+\dfrac{1}{v}\right)\left(\dfrac{m}{a}-\dfrac{1}{u}\right)\left(\dfrac{m}{a}+\dfrac{1}{u}\right)} \qquad (3.128)$$

when $n/b = 1/v$, $m/a = 1/u$

$$q_{mn} = 0 \qquad (3.129)$$

Of course any other lateral loading can be obtained through the use of (3.125).

If one now wishes to find the natural frequencies of this laminated composite plate, which has mid-surface symmetry ($B_{ij} = 0$), no other couplings, $(\ )_{16}, = (\ )_{26} = (\ )_{45} = 0$, but includes transverse shear deformation, $\epsilon_{xz} \neq 0$, $\epsilon_{yz} \neq 0$, then one sets $p(x, y) = 0$ in (3.115), but adds $-\rho h(\partial^2 w/\partial t^2)$ to the right hand side. In addition, because $\bar{\alpha}$ and $\bar{\beta}$ are both dependent variables which are independent of w, there will be an oscillatory motion of the lineal element across the plate thickness about the mid-surface of the plate which results in the right hand side of (3.113) and (3.114) becoming $I(\partial^2\bar{\alpha}/\partial t^2)$ and $I(\partial^2\bar{\beta}/\partial t^2)$ respectively, as shown below

$$
D_{11}\frac{\partial^2\bar{\alpha}}{\partial x^2} + D_{66}\frac{\partial^2\bar{\alpha}}{\partial y^2} + (D_{12} + D_{66})\frac{\partial^2\bar{\beta}}{\partial x \partial y} - 2A_{55}\left(\bar{\alpha} + \frac{\partial w}{\partial x}\right) = I\frac{\partial^2\bar{\alpha}}{\partial t^2}
$$

$$(3.129)$$

$$
(D_{12} + D_{66})\frac{\partial^2\bar{\alpha}}{\partial x \partial y} + D_{66}\frac{\partial^2\bar{\beta}}{\partial x^2} + D_{22}\frac{\partial^2\bar{\beta}}{\partial y^2} - 2A_{44}\left(\bar{\beta} + \frac{\partial w}{\partial y}\right) = I\frac{\partial^2\bar{\beta}}{\partial t^2}
$$

$$(3.130)$$

$$
2A_{55}\left(\frac{\partial\bar{\alpha}}{\partial x} + \frac{\partial^2 w}{\partial x^2}\right) + 2A_{44}\left(\frac{\partial\bar{\beta}}{\partial y} + \frac{\partial^2 w}{\partial y^2}\right) = \rho h\frac{\partial^2 w}{\partial t^2} \tag{3.131}
$$

where ρ is given by (3.85) and

$$
I = \rho h^3/12 \tag{3.132}
$$

Similar to the procedure used in Section 3.5, for the simply supported plate let

$$
w(x, y, t) = \sum_{m=1}^{\infty} \sum_{n=1}^{\infty} C'_{mn} \sin\frac{m\pi x}{a} \sin\frac{n\pi y}{b} e^{i\omega t} \tag{3.133}
$$

$$
\bar{\alpha}(x, y, t) = \sum_{m=1}^{\infty} \sum_{n=1}^{\infty} A'_{mn} \cos\frac{m\pi x}{a} \sin\frac{n\pi y}{b} e^{i\omega t} \tag{3.134}
$$

$$
\bar{\beta}(x, y, t) = \sum_{m=1}^{\infty} \sum_{n=1}^{\infty} B'_{mn} \sin\frac{m\pi x}{a} \cos\frac{n\pi y}{b} e^{i\omega t} \tag{3.135}
$$

Substituting these into the dynamic governing equations above results in a set of homogeneous equations which can be solved for the natural frequencies of vibration

$$
\begin{bmatrix}
L'_{11} & L_{12} & L_{13} \\
L_{12} & L'_{22} & L_{23} \\
L_{13} & L_{23} & L'_{33}
\end{bmatrix}
\begin{Bmatrix}
A'_{mn} \\
B'_{mn} \\
C'_{mn}
\end{Bmatrix}
=
\begin{Bmatrix}
0 \\
0 \\
0
\end{Bmatrix}
\tag{3.136}
$$

where the unprimed quantities were defined above and

$$
L'_{11} = L_{11} - \frac{\rho h^3}{12}\omega_{mn}^2
$$

$$
L'_{22} = L_{22} - \frac{\rho h^3}{12}\omega_{mn}^2
$$

$$
L'_{33} = L_{33} - \rho h \omega_{mn}^2
$$

As shown before [2] three eigenvalues (natural frequencies) result from solving (3.136) for each value of m and n. However, two of the frequencies are extremely higher than the other because they are associated with the rotatory inertia terms, that is, the right hand sides of (3.129) and (3.130) and are very seldom important in structural responses. If they are neglected then $L'_{11} = L_{11}$ and $L'_{22} = L_{22}$ above, and the square of the remaining natural frequency can be easily found to be

$$
\omega_{mn}^2 = \left[QL_{33} + 2L_{12}L_{23}L_{13} - L_{22}L_{13}^2 - L_{11}L_{23}^2 \right] / \rho h Q
\tag{3.137}
$$

where, here, $Q = L_{11}L_{22} - L_{12}^2$.
Also,

$$
A'_{mn} - \frac{L_{12}L_{23} - L_{22}L_{13}}{Q} C'_{mn}
$$

$$
B'_{mn} = \frac{L_{12}L_{13} - L_{11}L_{23}}{Q} C'_{mn}
\tag{3.138}
$$

If transverse shear deformation effects were neglected (3.137) would be identical to the square of (3.87).

Dobyns then goes on to develop the solutions for the simply supported laminated composite plate used throughout this Section for the plate subjected to a dynamic lateral load $p(x, y, t)$, neglecting the rotatory inertia terms discussed above, utilizing a convolution integral $P(t)$ as seen below: Incidentally the convolution integral is also known as the superposition integral and the Duhamel integral

$$w(x, y, t) = \frac{1}{\rho h} \sum_{m=1}^{\infty} \sum_{n=1}^{\infty} \left(\frac{q_{mn}}{\omega_{mn}} \right) \sin \frac{m\pi x}{a} \sin \frac{n\pi y}{b} P(t) \tag{3.138}$$

$$\bar{\alpha}(x, y, t)$$

$$= \frac{1}{\rho h} \sum_{m=1}^{\infty} \sum_{n=1}^{\infty} \left(\frac{q_{mn}}{\omega_{mn}} \right) \frac{(L_{12}L_{23} - L_{22}L_{13})}{Q} \cos \frac{m\pi x}{a} \sin \frac{n\pi y}{b} P(t)$$

$$\tag{3.139}$$

$$\bar{\beta}(x, y, t) = \frac{1}{\rho h} \sum_{m=1}^{\infty} \sum_{n=1}^{\infty} \left(\frac{q_{mn}}{\omega_{mn}} \right) \frac{L_{12}L_{13} - L_{11}L_{23}}{Q} \sin \frac{m\pi x}{a} \cos \frac{n\pi y}{b} P(t)$$

$$\tag{3.140}$$

where

$$P(t) = \int_0^t F(\tau) \sin \omega_{mn}(t - \tau) d\tau \tag{3.141}$$

and q_{mn} is the coefficient of the lateral load function expanded in series form, see (3.119).

So for a given lateral distributed load $p(x, y, t)$, if a solution of the form given by (3.138) through (3.140) is applicable, then the curvatures κ_x, κ_y, and κ_{xy} for the plate can be found through (3.99), and the stresses in each lamina can be found from (3.124).

The function $P(t)$ has been solved analytically for several representative forcing functions shown in Figure 3.9.

For the sine pulse the forcing function $F(t)$ and the convolution integral $P(t)$ are:

$$F(t) = F_0 \sin(\pi t/t_1) \quad 0 \leqslant t \leqslant t_1$$
$$F(t) = 0 \quad\quad\quad\quad t > t_1$$

$$P(t) = \int_0^t F(t) \sin \omega_{mn}(t - \tau) d\tau$$

$$= \frac{F_0 t_1 [\pi \sin \omega_{mn} t - \omega_{mn} t_1 \sin(\pi t/t_1)]}{(\pi^2 - t_1^2 \omega_{mn}^2)} \quad \text{for} \quad 0 \leqslant t \leqslant t_1 \tag{3.142}$$

$$P(t) = \frac{F_0 \pi t_1 [\sin \omega_{mn} t + \sin \omega_{mn}(t - t_1)]}{\pi^2 - t_1^2 \omega_{mn}^2} \quad \text{for} \quad t \geqslant t_1 \tag{3.143}$$

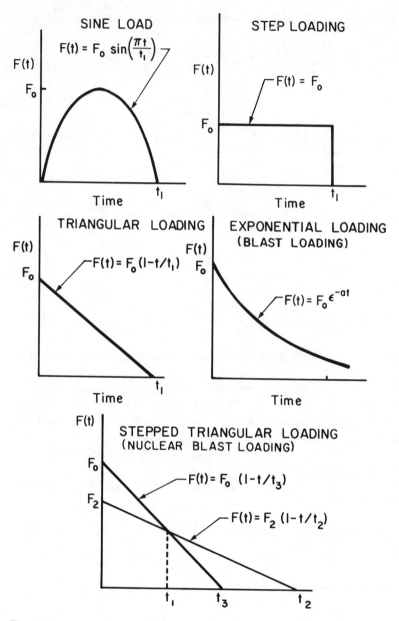

Figure 3.9.

For the stepped pulse the forcing functions $F(t)$ and the convolution integral $P(t)$ are given by:

$$F(t) = F_0 \quad 0 \leqslant t \leqslant t_1$$
$$F(t) = 0 \quad t > t_1$$

$$P(t) = \int_0^t F(\tau) \sin \omega_{mn}(t-\tau)d\tau = \frac{F_0}{\omega_{mn}}[1 - \cos \omega_{mn}t] \quad 0 \leqslant t \leqslant t_1$$

$$\text{(3.144)}$$

$$P(t) = \frac{F_0}{\omega_{mn}}[\cos \omega_{mn}(t-t_1) - \cos \omega_{mn}t] \quad t > t_1 \tag{3.145}$$

For a triangular pulse:

$$F(t) = F_0(1 - t/t_1) \quad 0 \leqslant t \leqslant t_1$$
$$F(t) = 0 \qquad\qquad t > t_1$$

$$P(t) = \int_0^t F(\tau) \sin \omega_{mn}(t-\tau)d\tau$$

$$= \frac{F_0}{\omega_{mn}}\left[1 - \cos \omega_{mn}t + \frac{1}{\omega_{mn}t_1}\sin \omega_{mn}t - t/t_1\right] \quad 0 \leqslant t \leqslant t_1 \quad \text{(3.146)}$$

$$P(t) = F_0\left[-\frac{1}{\omega_{mn}}\cos \omega_{mn}t\right.$$

$$\left. + \frac{2}{\omega_{mn}^2 t_1}\cos \omega_{mn}(t - t_{1/2}) \sin \omega_{mn}(t_{1/2})\right] \quad t > t_1 \tag{3.147}$$

The stepped triangular pulse of Figure 3.9 simulates a nuclear blast loading [10] where the pressure pulse consists of a long duration phase of several seconds due to the overpressure and a short duration phase of a few milliseconds due to the shock wave reflection. The short duration phase has twice the pressure of the long duration phase.

$$F(t) = F_0(1 - t/t_3) \quad 0 \leqslant t \leqslant t_1$$
$$F(t) = F_0(1 - t/t_2) \quad 0 \leqslant t \leqslant t_2$$
$$F(t) = 0 \qquad\qquad t > t_2$$

$$P(t) = \int_0^t F(\tau) \sin \omega_{mn}(t-\tau)d\tau$$

$$= \frac{F_0}{\omega_{mn}}\left[1 - \cos \omega_{mn}t + \frac{1}{\omega_{mn}t_3}\sin(\omega_{mn}t) - t/t_3\right] \quad 0 \leqslant t \leqslant t_1 \tag{3.148}$$

$$P(t) = F_0 \left[\frac{1}{\omega_{mn}} (1 - t/t_3) \cos \omega_{mn}(t - t_1) \right.$$

$$\left. - \frac{1}{\omega_{mn}} \cos \omega_{mn} t - \frac{1}{\omega_{mn}^2 t_3} \sin \omega_{mn}(t - t_1) + \frac{1}{\omega_{mn}^2 t_3} \sin \omega_{mn} t \right]$$

$$+ F_2 \left[\frac{1}{\omega_{mn}} (1 - t/t_2) - \frac{1}{\omega_{mn}} (1 - t/t_2) \cos \omega_{mn}(t - t_1) \right.$$

$$\left. + \frac{1}{\omega_{mn}^2 t_2} \sin \omega_{mn}(t - t_1) \right] \quad t_1 \leqslant t \leqslant t_2 \qquad (3.149)$$

$$P(t) = F_0 \left[\frac{1}{\omega_{mn}} (1 - t_1/t_3) \cos \omega_{mn}(t - t_1) \right.$$

$$\left. - \frac{1}{\omega_{mn}} \cos \omega_{mn} t + \frac{1}{\omega_{mn}^2 t_3} \{ \sin \omega_{mn} t - \sin \omega_{mn}(t - t_1) \} \right]$$

$$+ F_2 \left[\frac{1}{\omega_{mn}} (t_1/t_2 - 1) \cos \omega_{mn}(t - t_1) \right.$$

$$\left. - \frac{1}{\omega_{mn}^2 t_2} \{ \sin \omega_{mn}(t - t_2) - \sin \omega_{mn}(t - t_1) \} \right] \quad t > t_2 \qquad (3.150)$$

Lastly the exponential pulse of Figure 3.9 may be used to simulate a high explosive (non-nuclear) blast loading when the decay parameter α is empirically determined to fit the pressure pulse of the actual blast. The equations are

$$F(t) = F_0 e^{-\alpha t}$$

$$P(t) = \int_0^t F(\tau) \sin \omega_{mn}(t - \tau) \, d\tau$$

$$= \frac{F_0 \left[\omega_{mn} e^{-\alpha t} + \alpha \sin \omega_{mn} t - \omega_{mn} \cos \omega_{mn} t \right]}{(\alpha^2 + \omega_{mn}^2)} \quad t > 0 \qquad (3.151)$$

For the reader it should be noted that although the forcing function equations given above are herein used to investigate the dynamic response of a composite material plate, these equations are useful for many other purposes.

Several example problems were investigated by Dobyns [10] and appear in his paper, including the impact of an elastic spherical ball, which is a nonlinear problem too detailed to include herein. He also compared his closed form solutions to those obtained from the STAGS-C finite difference solutions.

Dobyns concludes that the equations presented in this section allow one to analyze a composite material panel subjected to dynamic loads with only a little more effort than is required for the same panel subjected to static loads. He concludes that he easily developed a Basic language program for a desktop computer to study the various dynamic loadings discussed herein. Hence, he concludes, one does not have to rely upon approximate design curves or arbitrary dynamic magnification load factors.

With the information presented herein, the necessary equations for the study of a composite plate without various couplings can be made, but including transverse shear deformation, subjected to various static lateral loads and a variety of dynamic loads which can be used singly – or superimposed – to describe a complex dynamic input. (Those same load functions can be used for beams, shells and many other structural configurations.) With the solutions for $\bar{\alpha}(x, y)$, $\bar{\beta}(x, y)$ and w, maximum deflections and stresses can be determined for deflection (stiffness) critical and strength critical structures.

For additional studies of composite plates, including transverse shear deformation effects, see Chapter 6 of Reference 2.

3.12. Some Remarks on Composite Structures

So far in this chapter, plates made of composite materials have been discussed. However, there are complicated constructions which are made either from composite materials or from isotropic materials which can be referred to as composite structures. A sandwich panel, the subject of the next section is one of these. Another is a box beam construction shown below in Figure 3.10, which could be the cross-section of a wind mill blade, a water ski, or many other structural components.

Such a structure will be subjected to tensile or compressive loads in the x-direction, to bending loads about the structural mid-surface, and to torsional loads about the x-axis. In each case one needs to develop the extensional stiffness matrix EA, the flexural stiffness matrix EI, and the torsional stiffness matrix GJ, for the rectangular cross-section.

It is probable that in the structural component considered, the top and bottom panels would be identical, as well as each side panel perpendicular to the other one – that will be assumed here, and therefore the subscripts 1 and 2 will be used.

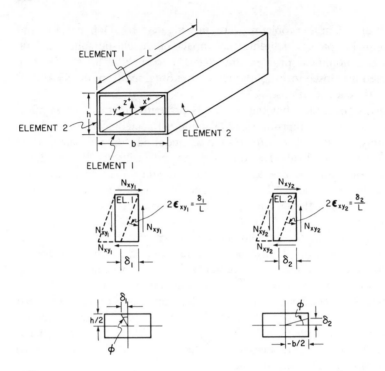

Figure 3.10.

For each panel the extensional relationship in the x-direction involves, for a construction without couplings

$$N_x = A_{11}\epsilon_x^0, \quad \text{where} \quad N_x \text{ is the force per unit width}$$

Simply adding the contribution of each unit width, the overall load P carried by the overall construction can be written as

$$P = 2N_{x_1}b + 2N_{x_2}h = \left[2(A_{11})_1 b + 2(A_{11})_2 h\right]\epsilon_x^0$$

Hence, the structural extensional stiffness EA for the rectangular construction of Figure 3.10 is simply

$$EA = 2(A_{11})_1 b + 2(A_{11})_2 h$$

Similarly if the box beam is bent in the x-z plane the overall bending moment M will be related to the overall curvature κ_x, by

$$M = \left[2(D_{11})_1 b + 2(A_{11})_1 b\left(\frac{h}{2}\right)^2 + \frac{2(A_{11})_2 h^3}{12}\right]\kappa_x \tag{3.152}$$

However, if the top and bottom surfaces are thin compared to the overall box height h, then the first term is negligible compared to the other ones, so

$$(EI)_{\text{box beam}} = 2(A_{11})_1 b(h/2)^2 + \frac{(A_{11})_2 h^3}{6} \tag{3.153}$$

Similar expressions can easily be constructed for the torsional stiffness. Consider the construction of Figure 3.9 subjected to a torsional load T in inch-lbs. about the x-axis.

Then it is clear that

$$T = 2(N_{xy_1}) b\left(\frac{h}{2}\right) + 2(N_{xy_2}) h\left(\frac{b}{2}\right) \tag{3.154}$$

Now from Chapter 2, for both elements,

$$N_{xy_i} = 2A_{66_i}\epsilon^0_{xy_i} \quad (i = 1, 2)$$

If ϕ is the angle of twist caused by the torque T over the length L, then for element 1 and 2

$$\phi = \frac{\delta_1}{(h/2)} \quad \text{and} \quad \phi = \frac{\delta_2}{b/2} \tag{3.155}$$

It is also seen that

$$\epsilon_{xy_1} = \frac{\delta_1}{2L} \quad \text{and} \quad \epsilon_{xy_2} = \frac{\delta_2}{2L} \tag{3.156}$$

$$T = 2(N_{xy})_1 b\frac{h}{2} + 2(N_{xy})_2 h\frac{b}{2} = 2bh(A_{66})_1\epsilon_{xy_1} + 2bh(A_{66})_2\epsilon_{xy_2}$$

$$= bh(A_{66})_1\delta_1 + bh(A_{66})_2\delta_2$$

$$T = \frac{bh}{2L}\left[(A_{66})_1 h + (A_{66})_2 b\right]\phi$$

So the GJ, the torsional stiffness of the construction of Figure 3.9 is

$$GJ = \frac{bh}{2}\left[(A_{66})_1 h + (A_{66})_2 b\right] \tag{3.157}$$

The above merely illustrates what one can and must do to develop the basic mechanics of materials gross formulation for the extension, bending or twisting of a rectangular section, perhaps composed of very esoteric

composite materials but used for a water ski, windmill blade or many
other shapes for many other purposes.

3.13. Methods of Analysis for Honeycomb Core Sandwich Panels with Composite Material Faces; and Their Structural Optimization

3.13.1. Introduction

A commonly used structural architecture is a honeycomb core sandwich construction, as shown in Figure 3.11 below.

The reason for its use is largely because of the great bending stiffness
resulting from the load carrying faces being separated by the core, the
same reason an I-beam has such great use.

For this construction shown there are several ways the structure may
fail when subjected to an in-plane compressive load $\overline{N}_x = -N_x$. One very
obvious way is if the honeycomb sandwich structure is overstressed. For
a honeycomb core sandwich panel one assumes, a priori, that the core
does not carry any appreciable load. If it did, it would simply bend or
buckle out of the way very near the ends and the load would never be
transmitted or resisted. Hence, the faces are loaded according to the
following equation:

$$\sigma_{F_x} = \frac{N_x}{2t_F} \qquad (3.158)$$

where N_x is the load per unit width of the plate, t_F is the face thickness
and σ_{F_x} is the stress in the face in the x-direction due to the applied load.

Over the above overstressing, in compression or tension, the plate can
buckle in various ways if the in-plane load is compressive. These failures
are addressed in the following paragraphs and include,
• Overall Elastic Instability
• Core Shear Instability
• Face Wrinkling Instability

3.13.2. Overall Elastic Instability

In this mode the overall plate buckles, similar to the buckling of the
composite plates discussed in earlier sections of this chapter. If the plate
is simply supported on all edges, this implies that a integer number of
half sine waves in the x-direction and one half of a sine wave in the
y-direction can be used to describe the phenomenon. For other boundary
conditions the buckling mode shapes vary, as they do in a simple plate,
composite or isotropic.

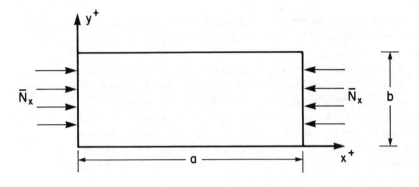

(a) Planview

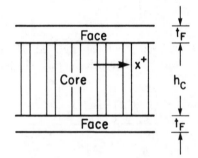

(b) Sideview

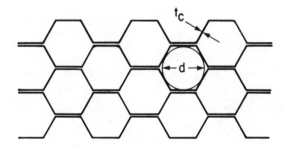

(c) Honeycomb core in planview

Figure 3.11.

In all of this Section the analysis is based upon that performed by Vinson and Shore [19], although they in turn reference many researchers, whose work they expanded upon.

Also, in what follows it is assumed that the composite faces are of such a construction that they can be considered as orthotropic plates, that is, the flexural stiffness of a face in the x-direction is D_x, in the y-direction D_y and in shear D_{xy}. The moduli of elasticity in the x and y directions are E_{f_x} and E_{f_y}. Here ν_{xy} and ν_{yx} are the Poisson's ratios of the faces, related by $E_{f_x}/\nu_{xy} = E_{f_y}/\nu_{yx}$. All of the laminated composite properties discussed previously can be related to these orthotropic properties if $B_{ij} = 0$, $(\ )_{16} = (\ )_{26} = (\ )_{45} = 0$.

For overall buckling of the plate, the critical stress can be written as:

$$\sigma_{cr} = -\frac{\pi^2}{4(1 - \nu_{xy}\nu_{yx})}\sqrt{E_{f_y}E_{f_x}}\left(\frac{h_c}{b}\right)^2 K_m \tag{3.159}$$

where all terms but K_m have been previously defined. The researchers at the Forest Products Laboratory as cited in Reference 21, define K_m to be

$$K_m = \frac{B_1 C_1 + 2B_2 C_2 + \dfrac{C_3}{B_1} + A\left[\dfrac{V_y}{C_4} + V_x\right]}{1 + (B_1 C_1 + B_3 C_2)\dfrac{V_y}{C_4} + \left(\dfrac{C_3}{B_1} + B_3 C_2\right)V_x + \dfrac{V_y V_x A}{C_4}} \tag{3.160}$$

where

$$A = C_1 C_3 - B_2 C_2^2 + B_3 C_2\left(B_1 C_1 + 2B_2 C_2 + \frac{C_3}{B_1}\right) \tag{3.161}$$

and

$$B_1 = \sqrt{D_y/D_x}$$

$$B_2 = \frac{D_y \nu_{xy}}{\sqrt{D_x D_y}} = 2B_3 + B_1 \nu_{xy}$$

$$B_3 = \frac{D_{xy}}{\sqrt{D_x D_y}}$$

$$V_x = \frac{\pi^2 \sqrt{D_x D_y}}{b^2 U_{xz}}$$

$$V_y = \frac{\pi^2 \sqrt{D_x D_y}}{b^2 U_{yz}}$$

$$U_{xz} = G'_{c_x} h_c$$

$$U_{yz} = G'_{c_y} h_c$$

G'_{c_x} = the effective honeycomb core transverse shear stiffness in the x-direction

$\quad = \frac{4}{3}(t_c/d)G_c$

G'_{c_y} = the effective honeycomb core transverse shear stiffness in the y-direction $= \frac{8}{15}(t_c/d)G_c$

G_c = Shear modulus of the core material itself.

Other constants in (3.160) are related to the boundary conditions, where of course integer values of n are sought to find the lowest values of K_m for the given geometry and materials involved.

All Edges Simply Supported

$$C_1 = C_4 = \frac{a^2}{n^2 b^2}; \quad C_2 = 1; \quad C_3 = \frac{n^2 b^2}{a^2}$$

Loaded Edges Simply Supported; Other Edges Clamped

$$C_1 = \frac{16}{3}\frac{a^2}{n^2 b^2}; \quad C_2 = 4/3; \quad C_3 = \frac{n^2 b^2}{a^2}; \quad C_4 = \frac{4}{3}\frac{a^2}{n^2 b^2}$$

Loaded Edges Clamped, Other Edges Simply Supported

$$C_1 = C_4 = \frac{3}{4}\frac{a^2}{b^2} \quad \text{for} \quad n = 1$$

$$C_1 = C_4 = \frac{1}{(n^2 + 1)}\frac{a^2}{b^2} \quad \text{for} \quad n \geqslant 2$$

$$C_2 = 1; \quad C_3 = \frac{n^4 + 6n^2 + 1}{(n^2 + 1)}\frac{b^2}{a^2}$$

All Edges Clamped

$$C_1 = 4C_4 = \frac{4a^2}{b^2} \quad \text{for} \quad n = 1$$

$$C_1 = 4C_4 = \frac{16}{3(n^2 + 1)}\frac{a^2}{b^2} \quad \text{for} \quad n \geqslant 2$$

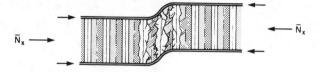

Figure 3.12.

$$C_2 = 4/3; \quad C_3 = \frac{n^4 + 6n^2 + 1}{(n^2 + 1)} \left(\frac{b^2}{a^2} \right)$$

Now, it turns out that overall buckling will be the lowest buckling mode of failure if and only if the σ_{cr} of (3.159) is equal to or less than (3.164) and if

$$V_x < \frac{5}{2} \sqrt{\frac{E_{f_y}}{E_{f_x}}} \, k_1 \qquad (3.162)$$

where under each of the four boundary conditions listed above, in order,

$$k_1 = 1, \ 3/4, \ 1, \ 3/4$$

Thus $k_1 = 1$ when unloaded edges are simply supported and $k_1 = 3/4$ when unloaded ($y = 0$, b) edges are clamped.

3.13.3 Core Shear Instability

It turns out that if $V_x \geqslant 5/2\sqrt{E_{f_y}/E_{f_x}} \, k_1$, then another form of buckling occurs at a load lower than necessary to cause overall buckling, called core shear instability which has a buckling mode shown by Figure 3.12.

This occurs when the face stiffness has been made so great that the core fails in transverse shear upon incipient buckling even with in-plane loads. This mode shape shown in Figure 3.12 usually occurs all across the panel in the y-direction.

This type of failure will occur if and only if the applied stress given in (3.158) reaches the critical value shown below and if $V > 5/2\sqrt{E_{f_y}/E_{f_x}} \, k_1$;

$$\sigma_{cr} = -\frac{G'_{c_x} h_c}{2t_F} = -\frac{2}{3} \left(\frac{t_c}{a} \right) \left(\frac{h_c}{t_F} \right) G_c \qquad (3.163)$$

3.13.4. Face Wrinkling Instability

A third kind of failure that can occur to a sandwich panel occurs when the face is just too thin, where both the total structure and the core have

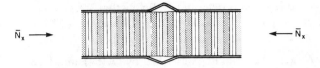

$\bar{N}_x \longrightarrow$ $\longleftarrow \bar{N}_x$

Figure 3.13.

much more resistance to buckling than the faces, considered as thin plates on an elastic foundation. Hence the faces fail as a subcomponent – but that results in total panel failure. In that case when the applied stresses of (3.158) attain a value of σ_{cr} shown below the panel fails

$$\sigma_{cr} = \left[\frac{16}{9} \left(\frac{t_F}{h_c} \right) \left(\frac{t_c}{a} \right) \frac{E_c \sqrt{E_{f_x} E_{f_y}}}{\left(1 - \nu_{xy} \nu_{yx} \right)} \right]^{1/2} \tag{3.164}$$

this kind of instability is shown graphically in Figure 3.13, where E_c is the elastic modulus of the core material in the z-direction.

Nothing much happens to the core during this kind of instability, but the faces invariably buckle outward with a very small wavelength all across the plate in the y-direction as stated above this results in the inability of the faces (the load carrying component) to carry the load – hence panel failure.

3.13.5. Monocell Buckling or Face Dimpling

Finally one other form of buckling can occur when the cell size of the honeycomb core is fairly large. The face buckles over one cell of the core, one can argue whether this could cause panel failure or not – so be it, but if that dimple tripped a boundary layer from laminar to turbulent flow with deleterious aerodynamic results or precipitated the detachment locally of the adhesive bond that resulted in the "unzipping" of the bond between the face and the core – then this could be the cause of panel failure. This type of buckling occurs when the stress of (3.158) attains a value equal to or greater than σ_{cr} given below:

$$\sigma_{cr} = -\frac{2\sqrt{E_{f_x} E_{f_y}}}{\left(1 - \nu_{xy} \nu_{yx} \right)} \left(\frac{t_F}{a} \right)^2 \tag{3.165}$$

3.13.6. Summary

It is seen that one way for the structure to fail is by overstressing – that is the stresses in the face given by (3.158) exceeding the allowable stress, however the allowable stress is determined. Below that stress, there are four different ways the panel can buckle: Overall buckling, (3.159), where $V_x \leqslant 5/2\sqrt{E_{f_y}/E_{f_x}}\,k_1$; core shear instability described by (3.163), where $V_x \geqslant 5/2\sqrt{E_{f_y}/E_{f_x}}\,k_1$; face wrinkling described by (3.164), and face dimpling given by (3.165). Curiously enough there are four geometric constants that one can vary for a given panel length and width, load, and material system; namely, t_F, h_c, t_c, and d. Four equations and four unknowns immediately suggests a unique solution – and sure enough one exists which results in a minimum weight structure for a given load per unit width, N_x. That process is called structural optimization.

3.13.7. Structural Optimization of the Honeycomb Core Sandwich Panel Subjected to an In-Plane Compressive Load

It follows then that with these four equations for buckling involving four geometric variables, that the "best" or "optimum" construction is when σ_{cr} for any buckling mode is the same as for all the others – otherwise, we can move material (weight) around to bring the "weakest link" up to the other buckling loads – because very surely the "weakest link" theory holds here, that is the panel fails whenever any of the four failure modes occurs.

Looking more closely we have the equations, a very involved and awkward one for overall "instability" which holds for $V \leqslant 5/2\sqrt{E_{f_y}/E_{f_x}}\,k_1$. Similarly a very easy to use equation for core shear instability (3.163) which holds for $V_x \leqslant 5/2\sqrt{E_{f_y}/E_{f_x}}\,k_1$. If we are going to equate all failure modes to the same buckling stress, then between (3.159), (3.163) and if both are equal then $V_x = 5/2\sqrt{E_{f_y}/E_{f_x}}\,k_1$, the easier two of the three to use are $V_x = 5/2\sqrt{E_{f_y}/E_{f_x}}\,k_1$ and (3.163) to express all three. Then, one can use 3.164 for face wrinkling and (3.165) for face dimpling. Therefore, these are the four equations and four unknowns: t_c, d, h_c, and t_F. It is advisable to use (3.158) to relate face stresses to applied load, and the weight per unit planform is given by:

$$(W - W_{ad}) = 2\rho_F t_F + \frac{8}{3}\left(\frac{t_c}{d}\right)h_c\rho_c \tag{3.166}$$

where ρ_F and ρ_c are the weight densities of the face and core material respectively, and W_{ad} is the weight of the adhesive material which makes a structural connection between core and face. This adhesive weight is a non-analytic weight per unit planform area which depends solely on the skill of the people fabricating the sandwich panel.

So for structural optimization we have six equations and six unknowns: h_c, t_F, t_c, d, σ_{f_x} and $(W - W_{ad})$. Finally it has been shown that for these panels if we assume the unloaded edges simply supported (that is $k_1 = 1$) compared to the unloaded edges clamped (that is $k_1 = 3/4$), at most the optimized weight is 7.45% greater, assume $k_1 = 1$ and optimize on that.

The governing equations to use in the optimization are:

$$\sigma_{cr} = -\frac{2}{3}\left(\frac{t_c}{d}\right)\left(\frac{h_c}{t_F}\right)G_c \tag{3.167}$$

From $V_x = (5/2)\sqrt{E_{f_y}/E_{f_x}} = (3/4)$, one obtains

$$\frac{\pi^2}{2(1 - \nu_{xy}\nu_{yx})}\frac{3}{4}\frac{t_F h_c d}{b^2 t_c}\frac{E_{f_x}}{G_c} = \frac{5}{2} \tag{3.168}$$

$$\sigma_{cr} = -\left[\frac{16}{9}\frac{1}{(1 - \nu_{xy}\nu_{yx})}\left(\frac{t_F}{n_c}\right)\left(\frac{t_c}{d}\right)E_c\sqrt{E_{f_x}E_{f_y}}\right]^{1/2} \tag{3.169}$$

$$\sigma_{cr} = -\frac{2\sqrt{E_{f_x}E_{f_y}}}{(1 - \nu_{xy}\nu_{yx})}\left(\frac{t_F}{d}\right)^2 \tag{3.170}$$

$$N_x = 2t_F\sigma_{f_x} \tag{3.171}$$

$$(W - W_{ad}) = 2\rho_F t_F + \frac{8}{3}\left(\frac{t_c}{d}\right)h_c\rho_c \tag{3.172}$$

Interestingly enough the above equations can be manipulated to provide a unique relationship between the "load index", (N_x/b), a concept devised decades ago by NASA-Langley (NACA) and the stress σ_{f_x}, involving only material properties. This is termed a "universal relationship" between load and resulting stress. To accomplish this substitute (3.168) into (3.169) and then substitute (3.171) therein and one obtains the following:

$$\frac{N_x}{b} = \frac{\sqrt{15}\,(1 - \nu_{xy}\nu_{yx})}{\pi}\left(\frac{G_c}{E_c}\right)^{1/2}\frac{\sigma_{f_x}^2}{E_{f_x}^{3/4}E_{f_y}^{1/4}} \tag{3.173}$$

If a load per unit width in tension of compression exceeds the allowable tensile or compressive strength of the face material determined by (3.158) then the panel will be overstressed. However, if (N_x/b) is less than that value, then the panel can only fail by buckling; and if one really wants to achieve minimum weight then – for any N_x/b, the σ_{x_f} given by (3.173) will be that results in minimum weight. If you choose h_c, t_c, d, or t_F such that the σ_{x_f} is greater or less than that given by (3.173) – the panel will weigh more.

If one uses (3.173) to obtain the minimum weight structure for a honeycomb core sandwich panel the dimensions of every component will be as follows:

$$\frac{h_c}{b} = \frac{\sqrt{10}}{\pi} \left(1 - \nu_{xy}\nu_{yx}\right)^{1/2} \left(\frac{\sigma_{f_x}}{E_{f_x}}\right)^{1/2} \tag{3.174}$$

$$\frac{d}{b} = \frac{1}{\pi} \left[\frac{15}{2} \left(1 - \nu_{xy}\nu_{yx}\right)\right]^{1/2} \left(\frac{G_c}{E_c}\right)^{1/2} \left(\frac{\sigma_{f_x}}{E_{f_x}}\right)^{1/2} \tag{3.175}$$

$$\frac{t_c}{b} = \frac{9\sqrt{5}}{8\pi} \left(1 - \nu_{xy}\nu_{yx}\right) \frac{\sigma_{f_x}^2}{E_c E_{f_x}^{3/4} E_{f_y}^{1/4}} \tag{3.176}$$

$$\frac{t_F}{b} = \frac{\sqrt{15}}{2} \frac{\left(1 - \nu_{xy}\nu_{yx}\right)}{\pi} \left(\frac{G_c}{E_c}\right)^{1/2} \frac{\sigma_{f_x}}{E_{f_x}^{3/4} E_{f_y}^{1/4}} \tag{3.177}$$

Calculations easily show that if the dimensions for h_c, d, t_c and t_F are used for any face stress $\sigma_{f_x} \leqslant \sigma_{all}$ work then all four buckling modes will occur simultaneously, and the minimum panel weight per unit area $(W - W_{ad})$ will be:

$$\frac{W - W_{ad}}{b} = \sqrt{15} \frac{\left(1 - \nu_{xy}\nu_{yx}\right)}{\pi} \left(\frac{G_c}{E_c}\right)^{1/2} \frac{\sigma_{f_x}}{E_{f_x}^{3/4} E_{f_y}^{1/4}} \left[\rho_F + 2\rho_c \frac{\sigma_{f_x}}{G_c}\right] \tag{3.178}$$

It is true that for a given N_x, material system, and geometry, it may be impractical to utilize the variables h_c, t_c, d, and t_F as optimumly determined. For instance, in commercially available honeycomb cores, usually d is smaller than optimum, and t_c is greater than optimum. This means that no commercially available core can be the cause of failure and thus h_c can be determined through machining. The geometrical quantity t_F can be determined to an almost optimum value because, commercially available composite tapes for making laminated faces are generally in the order of 0.007″ for graphite epoxy. At any rate to be safe, after determining the optimum dimensions, one is always safe in letting t_F, and t_c be slightly too large and h_c and d be slightly too small compared to the optimum value of (3.174) through (3.177), for a given materials system. Then one can compare the price paid for going with commercially available, but inexpensive, dimensions compared with the non-commercially available, but minimum weight, dimensions by comparing the finally chosen dimensions using (3.172) to the weight obtained optimally through (3.178). Thus a weight per unit money decision can be made, a truly engineering management decision.

3.13.8. Final Remarks

In a whole series of reports and papers the optimization of honeycomb core sandwich panels, square core honeycomb sandwich panels, and solid core sandwich panels with isotropic and composite material faces were analyzed and optimized for minimum weight, for in-plane compressive loads, lateral loads, in-plane shear loads and various combinations. In addition, web core sandwich panels and truss core sandwich panels were also analyzed and optimized [22, 23, 24.]

3.14. Concluding Remarks

It appears that there is no end in trying to more adequately describe mathematically the behavior of composite materials utilized in structural components. Unfortunately, the more sophisticated we get in such descriptions the more difficult the mathematics becomes, or is evidenced in the increasing difficulty observed as one progresses through the Sections of Chapter 3.

One additional complication that is important in *some* composite material structures is that the stiffness (and other properties) are different in tension than they are in compression. This occurs because (1) sometimes it is true of both fiber and matrix materials, but (2) sometimes it occurs because the matrix material is very weak compared to the fiber (that is $E_m \ll E_f$), so the fibers buckle in compression under small load so that for the composite the stiffness in compression differs markedly than that in tension. Hence, one can idealize a little and say that one has one set of elastic properties in tension and another set of elastic properties in compression. Bert has termed this a bimodulus material, typical of some composites, certainly typical of aramid (Kevlar) fibers in a rubber matrix as used in tires, and also typical of certain biological tissues modeled in biomedical engineering. In this context

$$\begin{bmatrix} A & B \\ \hline B & D \end{bmatrix}_{\text{Tension}} \neq \begin{bmatrix} A & B \\ \hline B & D \end{bmatrix}_{\text{compression}} \tag{3.179}$$

and all of the complications that result therefrom are too difficult to treat

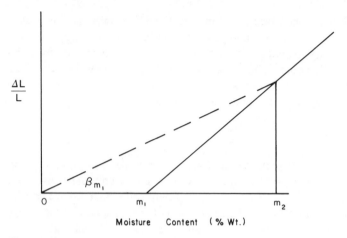

Figure 3.14.

in this text for first semester or first year students trying to learn the fundamentals of composite materials. However, one can refer to many references of Bert and co-workers [25,26,27], the original researcher in this area, and his excellent student Reddy [28].

Another complication is that of a more sophisticated description of hygrothermal effects. Crossman, Flaggs, and others have experimentally determined that for some composites $\beta = 0$ up to a moisture content m_1, then has a constant value for greater moisture content, as shown in Figure 3.14.

This most certainly introduces a non-linear aspect to all analysis, analogous to plasticity effects. However, again, if one is interested in steady state moisture effects at some moisture content, say m_2, usually 2%, then one could say that the coefficient of moisture expansion β is a value of β_{M_1}, and proceed with the linear analysis of earlier Sections. Certainly such an approximation is acceptable for preliminary design and analysis.

Lastly, time dependent effects in the stresses, deformation and strains of composite materials are becoming more important design considerations. Viscoelasticity and creep are respected disciplines about which entire books have been written. These effects have been deemed important in some composite material structures. Crossman, Flaggs, Vinson and Wilson have all commented thereupon. Wilson and Vinson [29] have shown that the effects of viscoelasticity on the buckling resistance of polymer matrix composite material plates is very significant. Similarly, the effect of viscoelasticity on the natural vibration frequencies will also be significant. Many of these effects have been included in a recent survey article by Reddy [30] who has focused primarily on plates of composite materials.

3.15. Problems and Exercises

3.1. Find the critical buckling load, $N_{x_{cr}}$, in lbs/in, for the plate when it is simply supported on all edges whose properties are given as:

$$D_{11} = 1.63 \times 10^6 \quad \text{lb-in} \quad \rho = 1.8 \times 10^{-4} \frac{\text{lb-sec}^2}{\text{in}^4}$$

$$D_{12} = 0.028 \times 10^6 \quad \text{lb-in} \quad a = 30''$$

$$D_{22} = 0.160 \times 10^6 \quad \text{lb-in} \quad b = 20''$$

$$D_{66} = 0.037 \times 10^6 \quad \text{lb-in} \quad h = 1''$$

3.2. Find the fundamental natural frequently in cps. for the plate of problem 3.2 simply supported on all edges.

3.3. The following material properties are given for a unidirectional, 4 ply laminate, $h = 0.020''$

$$A = \begin{vmatrix} 0.84 & 0.00547 & 0 \\ 0.00547 & 0.0176 & 0 \\ 0 & 0 & 2 \times 0.012 \end{vmatrix} \times 10^6 \; lb/\text{in}$$

where $A_{66} = 0.012 \times 10^6$ lb/in

$B = 0$

$$D = \begin{vmatrix} 28.053 & 0.1824 & 0 \\ 0.1824 & 0.5879 & 0 \\ 0 & 0 & 2 \times 0.4 \end{vmatrix} \text{lb-in}$$

where $D_{66} = 0.4$ lb-in

ρ, the mass density (corresponding to 0.06 lb/in^3) $= 1.554 \times 10^{-4}$ lb sec^2/in^4

Consider a plate made of the above material with dimensions $a = 20''$, $b = 30''$, $h = 0.020''$. For the *first* perturbation method of Section 3.8 determine $\bar{b}$ and α. Is α a proper value to use this perturbation technique?

3.4. Consider the plate of problem 3.3. If it is simply supported on all four edges, what is its fundamental natural frequency in cycles per seconds neglecting transverse shear deformation?

3.16. Bibliography

1. Vinson, J.R. *Structural Mechanics: The Behavior of Plates and Shells.* New York: Wiley-Interscience, John Wiley and Sons, 1974.
2. Vinson, J.R. and Chou, T.W. *Composite Materials and Their Use in Structures.* London: Applied Science Publishers, 1975.
3. Levy, M., "Sur L'equilibrie Elastique d'une Plaque Rectangulaire," *Compt Rend 129* (1899): 535–539.
4. Vinson, J.R., "New Techniques of Solutions for Problems in Orthotropic Plates," Ph.D. Dissertation, University of Pennsylvania, 1961.
5. Vinson, J.R. and Brull, M.A., "New Techniques of Solutions of Problems in Orthotropic Plates," *Transactions of the Fourth United States Congress of Applied Mechanics,* 1962.
6. Thomson, W.T. *Vibration Theory and Applications, 2nd Edition,* Prentice Hall, Inc., (1981).
7. Warburton, G., "The Vibration of Rectangular Plates," *Proceedings of the Institute of Mechanical Engineers,* 1968 (1954), 371.
8. Young, D. and Felgar, R., Jr., "Tables of Characteristic Functions Representing Normal Modes of Vibration of a Beam," *The University of Texas Publication Number 4913,* (1944).
9. Felgar, R., Jr., "Formulas for Integral Containing Characteristic Functions of a Vibrating Beam," *Bureau of Engineering Research, The University of Texas Publication,* 1950.
10. Dobyns, A.L., "The Analysis of Simply-Supported Orthotropic Plates Subjected to Static and Dynamic Loads," *American Institute of Aeronautics and Astronautics Journal,* (May 1981): 642–650.
11. Sloan, J.G. and Vinson, J.R., "The Behavior of Rectangular Composite Material Plates Under Lateral and Hygrothermal Loads," *AFOSR TR-78-1477,* (July 1978).
12. Pipes, R.B., Vinson, J.R. and Chou, T.W., "On the Hygrothermal Response of Laminated Composite Systems," *Journal of Composite Materials,* Vol. 3, (April 1976).
13. Whitney, J.M., "The Effect of Transverse Shear Deformation on the Bending of Laminated Plates," *Journal of Composite Materials,* Vol. 3, (July 1969).
14. Wu, C.I. and Vinson, J.R., "Nonlinear Oscillations of Laminated Specifically Orthotropic Plates with Clamped and Simply Supported Edges," *Journal of the Acoustical Society of America, 49 5, Pt. 2,* (May 1971): 1561–1567.
15. Flaggs, D.L. *Elastic Stability of Generally Laminated Composite Plates Including Hygrothermal Effects,* MMAE Thesis, University of Delaware, (June 1978).
16. Smith, A.P., Jr., *The Effect of Transverse Shear Deformation on the Elastic Stability of Orthotropic Plates Due to In-Plane Loads,* MMAE Thesis, University of Delaware, (May 1973).
17. Linsenmann, D.R. *Stability of Plates of Composite Materials,* MMAE Thesis, University of Delaware, (June 1974).
18. Shen, C. and Springer, G.S., "Moisture Absorption and Desorption of Composite Materials," *Journal of Composite Materials,* vol. 10, (January 1976).
19. Vinson, J.R. and Shore, S., "Methods of Structural Optimization for Flat Sandwich Panels," *U.S. Naval Air Engineering Center Technical Report NAEC-ASL-1083,* (1965).
20. Vinson, J.R. and Shore, S., "Design Procedures for the Structural Optimization of Flat Sandwich Panels," *U.S. Naval Air Engineering Center Technical Report NAEC-ASL-1084,* (1965).
21. Vinson, J.R. and Shore, S., "Bibliography on Methods of Structural Optimization of Flat Sandwich Panels," *U.S. Naval Air Engineering Center Technical Report NAEC-ASl-1082,* (1965).

22. Vinson, J.R. and Shore, S., "Structural Optimization of Corrugated Core and Web-Core Sandwich Panels Subjected to Uniaxial Compression," *U.S. Naval Air Engineering Center Technical Report NAEC-ASL-1109*, (1967).

23. Vinson, J.R. and Shore, S., "Structural Optimization of Flat, Corrugated Core and Web-Core Sandwich Panels Under In-Plane Shear Loads, and Combined Uniaxial Compressive and In-Plane Shear Loads," *U.S. Naval Air Engineering Center Technical Report NAEC-ASL-1110*, (1967).

24. Vinson, J.R. and Shore, S., "A Method for Weight Optimization of Flat Truss-Core Sandwich Panels Under Lateral Loads," *U.S. Naval Air Engineering Center Technical Report NAEC-ASL-1111*, (1967).

25. Bert, C.W., "Vibration of Composite Structures," *U.S. Office of Naval Research Technical Report OU-AMNE-80-6*, (March 1980).

26. Bert, C.W., Reddy, J.N., Reddy, V.S. and Chao, W.C., "Analysis of Thick Rectangular Plates Laminated of Bimodulus Composite Materials," *AIAA Paper*, (May 1980).

27. Reddy, J.N. and Bert, C.W., "On the Behavior of Plates Laminated of Bimodulus Composite Materials," *VPI & SU Report VPI-E-81-11 and OU-AMNE-81-1*, (April 1981).

28. Reddy, J.N., "Analysis of Layered Composite Plates Accounting for Large Deflections and Transverse Shear Strains," *Recent Advances in Nonlinear Computational Mechanics*, Pine Ridge Press (1981).

29. Wilson, D.W. and Vinson, J.R., "Viscoelastic Analysis of Laminated Plate Buckling", AIAA Journal, Vol. 22, No. 7, July 1984, pp 982–988.

30. Reddy, J.N., "Survey of Recent Research in the Analysis of Composite Plates," *Composite Technology Review*, Vol. pp. (Fall 1982).

4. Beams, Columns and Rods of Composite Materials

4.1. Development of a Simple Theory

A beam, column or rod is a long thin structural component of width b, height h and length L, where $b/L \ll 1$ and $h/L \ll 1$, as shown in Figure 4.1.

The term beam is utilized when the structure of Figure 4.1 is subjected to a lateral load in the x-z plane, in the z-direction, distributed or concentrated, applied to the upper or lower surface (that is, $z = \pm h/2$), such that bending (curvature in the x, z plane) occurs. The term "rod" is used when the structure of Figure 4.1 is loaded in the axial direction (the x-direction) by tensile forces which try to "stretch" the structure. The term "column" is used when the structure shown is subjected to compres-

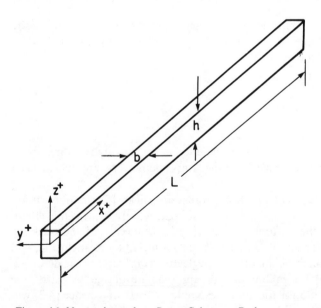

Figure 4.1. Nomenclature for a Beam, Column or Rod.

sive forces in the x-direction which reduces the length of the structure, resulting in compressive stresses, and/or the elastic instability (buckling) of the structure, a topic which will be discussed later in this chapter. Combination of these loads may occur, such as when the first and third load-types occur simultaneously, resulting in the structure being referred to as a beam column.

As a simple first example, consider the beam to be loaded in the x-z plane only, whether the loading is lateral, in-plane tensile, in-plane compressive or combinations thereof. Also, for simplicity, ignore the thermal and moisture effects (that is, hygrothermal effects) which add on to the strain terms in the constitutive equations given in Chapter 2. Because a beam is so narrow ($b \ll L$), strains are ignored in the y-direction, implying that all Poisson's ratio effects can be ignored (that is a classical beam assumption). Lastly, there is no y-direction dependence on any quantity involved in the set of governing equations.

Looking then at (2.62), the above assumptions dictate that the remaining constitutive equations for this beam are

$$
\begin{bmatrix} N_x \\ M_x \end{bmatrix} = \begin{bmatrix} A_{11} & B_{11} \\ B_{11} & D_{11} \end{bmatrix} \begin{bmatrix} \epsilon_x^0 \\ \kappa_x \end{bmatrix} \tag{4.1}
$$

and from (2.59)

$$
Q_x = 2A_{55}\epsilon_{xz} \tag{4.2}
$$

If the beam has mid-plane symmetry, as the majority of composite constructions do, then $B_{ij} = 0$, hence $B_{11} = 0$, and the equations above in (4.1) become uncoupled, i.e.,

$$
N_x = A_{11}\epsilon_x^0 \tag{4.3}
$$

$$
M_x = D_{11}\kappa_x \tag{4.4}
$$

Note that without B_{11}, the in-plane load, N_x, and its resulting strain ϵ_x^0 are completely uncoupled from the stress couple, M_x, and its resulting curvature, κ_x.

In most composite constructions, transverse shear deformation cannot be ignored (that is, $\epsilon_{xz} \neq 0$), because it significantly affects the lateral deformation, w; the natural frequencies of vibration; and the buckling loads. Yet, stress levels are only moderately affected by its inclusion, and these simplified analytical methods, neglecting transverse shear deformation can be used for preliminary or approximate design to "size" the structure initally. So for this initial example, $\epsilon_{xz} = 0$, and hence (4.2) will be ignored in what follows.

Looking now to the plate equilibrium equations of the previous chapter, from the beam assumptions made, it is seen that $N_y = N_{xy} = M_y = M_{xy} = Q_y = \tau_{1y} = \tau_{2y} = 0$; hence the remaining equilibrium equations become:

$$\frac{dN_x}{dx} + \tau_{1x} - \tau_{1y} = 0 \tag{4.5}$$

$$\frac{dQ_x}{dx} + p_1 - p_2 = 0 \tag{4.6}$$

$$\frac{dM_x}{dx} - Q_x + \frac{h}{2}[\tau_{2x} + \tau_{2y}] = 0 \tag{4.7}$$

Again, for simplicity of example, assume the surface shear stresses τ_{1x} and τ_{2x} are zero. Also it is seen that the beam will only react to the difference between p_1 and p_2 the normal surface traction on the top and bottom surface, hence:

$$p(x) = p_1 - p_2. \tag{4.8}$$

The result is that (4.5) through (4.8) become

$$\frac{dN_x}{dx} = 0 \tag{4.9}$$

$$\frac{dQ_x}{dx} + p(x) = 0 \tag{4.10}$$

$$\frac{dM_x}{dx} - Q_x = 0 \tag{4.11}$$

It should be remembered that N_x, Q_x and M_x are each plate type quantities which are per unit edge distance in the y-direction, and hence apply directly to a beam of "unit" width. However, since nothing varies in the y-direction for the beam in question, it is both traditional and easy to multiply all of the above equations by the beam width b, hence:

$$P \equiv N_x b, \; V \equiv Q_x b, \; M_b \equiv M_x b, \; q(x) \equiv p(x)b \tag{4.12}$$

Therefore, the governing beam equations for a beam of composite materials, with mid-plane symmetry in its stacking sequence, subjected to lateral and in-plane loads, ignoring hygrothermal effects and transverse shear deformation are:

$$P = bA_{11}\epsilon_x^0 = bA_{11}\frac{du_0}{dx} \tag{4.13}$$

$$M_x = bD_{11}\kappa_x = -bD_{11}\frac{d^2 w}{dx^2} \qquad (4.14)$$

$$\frac{dP}{dx} = 0 \qquad (4.15)$$

$$\frac{dV}{dx} + q(x) = 0 \qquad (4.16)$$

$$\frac{dM_b}{dx} - V = 0 \qquad (4.17)$$

From (4.15) it is seen that $P = $ constant (as can be seen from a simple free-body diagram of a beam), and therefore integration of (4.13) gives the relation for the mid-surface in-plane displacement, u_0:

$$u_0(x) = \left(\frac{P}{bA_{11}}\right)x + C_0 \qquad (4.18)$$

wherein C_0 is a constant of integration determined by where $u_0(x)$ is specified. If the rod is loaded by a tensile axial load P only, (4.18) provides the displacement, from which all stresses in every ply can be determined by

$$[\sigma_x]_k = [\overline{Q}_{11}]_k [\epsilon_x^0] = [\overline{Q}_{11}]_k \left(\frac{du_0}{dx}\right) \qquad (4.19)$$

If P is a compressive load, the same applies, except if the load P that would cause buckling is sought, a more refined theory is needed, which will be discussed later.

Analogously, substituting (4.14) into (4.17) and the result into (4.16) results in the following differential equation for the bending of the beam:

$$bD_{11}\frac{d^4 w}{dx^4} = q(x) \qquad (4.20)$$

This is the governing differential equation for a composite material beam with $B_{11} = \Delta T = \Delta m = \epsilon_{xz} = 0$. Four straightforward integrations of (4.20) yields the entire solution including four constants of integration with which to satisfy the boundary conditions. It can easily be shown that under these conditions if the beam utilized a one layer, isotropic material, then $bA_{11} = EA = Ebh$ and $bD_{11} = EI = Ebh^3/12$, for a beam of rectangular cross-section.

Once the solution of $w(x)$ is found from (4.20), the stresses in each of the laminae are found by

$$[\sigma_x]_k = z[\bar{Q}_{11}]_k[\kappa_x] = -[\bar{Q}_{11}]_k z \frac{d^2 w}{dx^2} \tag{4.21}$$

Obviously if both in-plane and lateral loads occur simultaneously, then the stresses in each laminae are found by the sum of (4.19) and (4.21)

$$[\sigma_x]_k = [\bar{Q}_{11}]_k \frac{dU_0}{dx} - [\bar{Q}_{11}]_k z \frac{d^2 w}{dx^2} \tag{4.22}$$

4.2. Some Simplified Composite Beam Solutions

From the theory for a composite beam-rod-column developed in Section 4.1, solutions to all such problems can be found directly.

For the bending of a beam, the solution of (4.20) results in four constants of integration which are used to solve boundary conditions at each end of the beam. Discussion of classical boundary conditions are found in any undergraduate mechanics textbook and so will not be developed herein. There are three classical boundary conditions: simple-support, clamped and free. For each, the explicit boundary conditions are:

Simple Support: Clamped: Free:

$w = 0$ $w = 0$ $M_b = 0$

$M_b = 0$ $\dfrac{dw}{dx} = 0$ $V_b = 0$ (4.23)

As an example, consider a beam clamped at each end subjected to a uniform lateral load $q(x) = -q_0$, where q_0 is a constant as shown in Figure 4.2.

From (4.20)

$$\frac{d^4 w}{dx^4} = -\frac{q_0}{bD_{11}}$$

$$\frac{d^3 w}{dx^3} = \left(-\frac{q_0}{bD_{11}}\right)x + C_1 \tag{4.24}$$

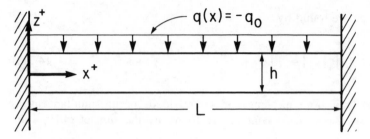

Figure 4.2. Clamped – Clamped beam with a uniform lateral load.

$$\frac{d^2 w}{dx^2} = -\frac{q_0}{bD_{11}}\frac{x^2}{2} + C_1 x + C_2 \tag{4.25}$$

$$\frac{dw}{dx} = -\frac{q_0}{bD_{11}}\frac{x^3}{6} + \frac{C_1 x^2}{2} + C_2 x + C_3 \tag{4.26}$$

$$w(x) = -\frac{q_0}{bD_{11}}\frac{x^4}{24} + \frac{C_1 x^3}{6} + \frac{C_2 x^2}{2} + C_3 x + C_4 \tag{4.27}$$

Note that for any lateral load in general say $q(x)$, the first term of (4.27) would be $(1/bD_{11}) \iiint q(x)\, dx dx dx dx$. The remaining terms would be the same.

The clamped boundary conditions are used now to solve for the constants C_1 through C_4.

$$w(0) = 0 = C_4$$

$$\frac{dw}{dx}(0) = 0 = C_3$$

$$w(L) = 0 = -\frac{q_0 L^4}{24bD_{11}} + \frac{C_1 L^3}{6} + \frac{C_2 L^2}{2}$$

$$\frac{dw}{dx}(L) = 0 = -\frac{q_0 L^3}{6bD_{11}} + \frac{C_1 L^2}{2} + C_2 L$$

Therefore,

$$C_1 = \frac{q_0 L}{2bD_{11}} \quad \text{and} \quad C_2 = -\frac{q_0 L^2}{12bD_{11}}$$

Finally,

$$w(x) = -\frac{q_0}{24bD_{11}}[x^4 - 2Lx^3 + L^2 x^2] \tag{4.28}$$

One objective of such analysis is to determine the maximum deflection, because some structures are deflection or stiffness critical. Therefore, it is seen (in this case through symmetry) that

$$w_{\max} = w(L/2) = -\frac{q_0 L^4}{384 b D_{11}} \tag{4.29}$$

Similarly, it must be determined where the maximum stress is in order to calculate that value because quite often structures are strength critical. From (4.14) and (4.28),

$$M_b = -b D_{11} \frac{d^2 w}{dx^2} = \frac{q_0}{12} [6x^2 - 6Lx + L^2]$$

Note that the bending moment M_b is not dependent upon any elastic material property. It is seen that the maximum value occurs at $x = 0$ and L, and is

$$M_{b_{\max}} = M_b(0, L) = \frac{q_0 L^2}{12} \tag{4.30}$$

Suppose the beam were of a one layer, isotropic material then the maximum stress would be at the top and bottom of the beam at each end, that is, $z = \pm h/2$, and the traditional equation would be used

$$\sigma_x = \frac{M_b z}{I} \text{ where } I = bh^3/12$$

So

$$\sigma_{x_{\max}} = \sigma_x \begin{pmatrix} 0 \\ L \end{pmatrix}, \pm h/2 \end{pmatrix} = \pm \frac{6 M_{b_{\max}}}{bh^2} = \pm \frac{q_0 L^2}{2 bh^2}$$

from which the maximum stress can be calculated, and compared to the allowable strength properties for the material.

However, for a beam of a given composite material, the calculation is not so simple. Having found $M_{b_{\max}}$ through (4.30), (4.14) must be utilized to find the maximum curvature $\kappa_{x_{\max}}$,

$$\kappa_{x_{\max}} = \kappa_x \begin{pmatrix} 0 \\ L \end{pmatrix} = \frac{M_{b_{\max}}}{b D_{11}} = +\frac{q_0 L^2}{12 b D_{11}} = \left(-\frac{d^2 w}{dx^2} \right) \tag{4.31}$$

Then, and only then can the maximum stress be calculated for each lamina through (4.21), namely

$$[\sigma_x]_{k_{\max}} = [\overline{Q}_{11}]_k [\kappa_{x_{\max}}] z = -[\overline{Q}_{11}]_k \left(\frac{d^2 w}{dx^2} \right)_{\max} z \tag{4.32}$$

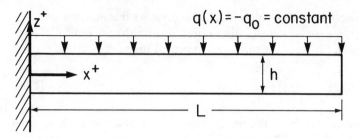

Figure 4.3. Clamped – Free Beam Subjected to a Uniform Load.

Then, this $[\sigma_x]_{k_{max}}$ must be compared with the allowable strength values (that is, failure stress) for each lamina with its specific orientation and composite material system.

Keep in mind that all of the above analysis could be more complicated even for a beam, if the beam were not mid-plane symmetric ($B_{11} \neq 0$) or if hygrothermal effects were present (that is, $\epsilon_x^0 + z\kappa_x - \alpha_x \Delta T - \beta_x \Delta m$), all of which were discussed in Chapter 2 in general.

As a second example of solutions to beam problems of composite materials consider a beam cantilevered from the end $x = 0$ as shown in Figure 4.3.

Again, the solution is (see 4.27),

$$w(x) = -\frac{q_0 x^4}{24bD_{11}} + \frac{C_1 x^3}{6} + \frac{C_2 x^2}{2} + C_3 x + C_4. \qquad (4.33)$$

Again, as before, because $w(0) = 0$ and $dw/dx(0) = 0$,

$$C_3 = C_4 = 0$$

The other two boundary conditions are $M_b(L) = 0$ and $V(L) = 0$, which is analogous to saying

$$\frac{d^2 w}{dx^2}(L) = 0 = -\frac{q_0 L^2}{2bD_{11}} + C_1 L + C_2$$

$$\frac{d^3 w}{dx^3}(L) = 0 = -\frac{q_0 L}{bD_{11}} + C_1$$

Hence:

$$C_1 = \frac{q_0 L}{bD_{11}} \quad \text{and} \quad C_2 = -\frac{q_0 L^2}{2bD_{11}}$$

Thus:

$$w(x) = -\frac{q_0}{24bD_{11}}[x^4 - 4Lx^3 + 6L^2x^2] \tag{4.34}$$

It is easily seen that again for maximum deflection and maximum stress couple (bending moment) that

$$w_{\max} = w(L) = -\frac{q_0 L^4}{8bD_{11}} \tag{4.35}$$

$$M_{b_{\max}} = M_b(0) = \frac{q_0 L^2}{2} \tag{4.36}$$

If this beam were of a one ply, isotropic material

$$\sigma_{x_{\max}} = \pm \frac{6M_{b_{\max}}}{bh^2} = \pm \frac{3q_0 L^2}{bh^2} \tag{4.37}$$

But, if the beam is made of a laminated composite material then from (4.21)

$$\kappa_{\max} = \kappa_x(0) = \frac{M_b}{bD_{11}} = \frac{q_0 L^2}{2bD_{11}} \tag{4.38}$$

and again

$$\sigma_{x_{\max_k}} = [\bar{Q}_{11}]_k z \kappa_{x_{\max}} \tag{4.39}$$

Many more examples for simplified composite beam-rod-columns (without buckling) could be given but they would be repetitive.

4.3. Bending of Laminated Beams – Advanced Theory.

4.3.1. Introductory Remarks

Due to the very nature of fabrication techniques used in advanced composite structures, an identification with the analysis and synthesis of plate and shell type structural elements appear quite natural. Therefore, not surprisingly, less attention has been focused on developing appropriate solution methodology for the more simplistic beam type structural components. For this latter case the general approach to developing solutions is based upon (a) Oversimplification of mechanics of material approaches to developing appropriate methodology, (b) Modifying existing plate theory by considering laminated strips for generating solution techniques, and (c) Attacking the problem from a more sophisticated viewpoint, that is, an anisotropic elasticity approach.

4.3.2. Basic Assumptions

In this section the basic governing equations pertaining to the hygrothermal mechanical behavior of a laminated straight beam are developed. The beam is considered to be made up from a number of elastic layers each of which possess different hygrothermomechanical properties, thicknesses, and geometric orientations.

The basic bending theory is based upon the following assumptions:
- Individual plies/laminae are integrally bonded (no slip)
- The thickness dimension of the beam is small relative to beam length
- Bernoulli-Euler Bending Theory is approximately applicable. Plane sections remain plane upon bending. This is not generally true for anisotropic material for which transverse normal and shear deformation may be important.
- Displacements, Rotations, and Strains are small relative to beam thickness
- A balanced ply orientation is used to preclude coupling effects between extension, flexure, and torsion
- Static loading only is considered. Inertia effects are omitted.
- Hygro and Thermal Effects are such that Creep effects are ignored.

Basic Equations (Classical)

The stress at any point in the beam is given by the following stress – strain law,

$$\sigma_x^{(k)} = E_x^{(k)} \epsilon_x^{(k)} \tag{4.40}$$

The corresponding beam stiffness coefficients are defined using the theory advanced in Chapter 2, that is

$$\{ A_x, B_x, D_x \} = \sum_{k=1}^{N} E_x^{(k)} t_k \left\{ h_k - h_{k-1}, \tfrac{1}{2}\left(h_k^2 - h_{k-1}^2 \right), \tfrac{1}{3}\left(h_k^3 - h_{k-1}^3 \right) \right\} \tag{4.41}$$

It should be noted that in the above expression the value of Young's modulus $E_x^{(k)}$ represents the beam stiffness along the reference geometric direction associated with the kth lamina. For cases in which the kth lamina is orthotropic and does not coincide with the beam's natural axis then the following transformation relation must be used to determine $E_x^{(k)}$,

$$\frac{1}{E_x^k} = \frac{\cos^4\theta_i}{E_{11}} + \left(\frac{1}{G_{12}} - \frac{2V_{12}}{E_{11}}\right)\cos^2\theta_k \sin^2\theta_k + \frac{\sin^4\theta_k}{E_{22}} \quad (4.42)$$

The displacement field for the beam including transverse normal and shear deformations and higher order terms, take the form *,

$$u = u^0 + z\psi_x + \frac{z^2}{2}\phi_x \quad (4.43)$$

$$w = w^0 + z\psi_z \quad (4.44)$$

For the classical problem, the strain field can be structured as shown below

$$\epsilon^{(k)}_{x_{\text{TOTAL}}} = \epsilon^{(k)}_{x_{\text{Mech}}} + \epsilon^{(k)}_{x_{\text{Ther}}} + \epsilon^{(k)}_{x_{\text{Hygr}}} \quad (4.45)$$

$$\epsilon^{(k)}_{x_{\text{TOTAL}}} = \epsilon^0_{x_{\text{TOTAL}}} + z\kappa_{x_{\text{TOTAL}}} \quad (4.46)$$

$$\text{where}\begin{cases} \epsilon^{(k)}_{x_{\text{Mech}}} = & \text{(Mechanical Loads)} \quad (4.47) \\ \epsilon^{(k)}_{x_T} = \alpha_x^{(k)}\Delta T & \text{(Thermal Loads)} \quad (4.48) \\ \epsilon^k_{x_H} = \beta_x^{(k)}\Delta m & \text{(Hygrometric Loads)} \quad (4.49) \end{cases}$$

In order to determine the appropriate stress equilibrium and stress resultant equations resort to the following laminated beam element as shown in Figure 4.4 is useful. It should be noted that since transverse shear deformation is suppressed in the initial calculation procedure, the method used to evaluate the transverse shear stress is that used in strength of materials techniques, that is,

$$\tau_{xz}^{(k)} = \frac{1}{t}\int_{n_1}^{z}\sigma_{x,x}^{(k)}\mathrm{d}A \quad (4.50)$$

It is possible to express the stress $\sigma_x^{(k)}$ over each lamina thickness and the summation over n laminae as,

$$N_{x_{\text{Mech}}*} = \sum_{k=1}^{N}\int_{h_{k-1}}^{h_k}\sigma_x^{(k)}\mathrm{d}A = A_x\epsilon^0_{x_{\text{TOTAL}}} + B_x\kappa_{x_{\text{TOTAL}}} - N_{x_T} - N_{x_H} \quad (4.51)$$

* It should be noted that in the Classical Theory the underscored terms representing coupling between in-plane and vertical displacements as well as higher order shear corrections are generally omitted.

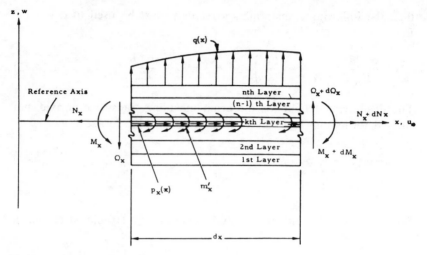

Figure 4.4. Laminated Beam Element.

$$M_{x_{\text{Mech}}} = \sum_{k=1}^{N} \int_{h_{k-1}}^{h_K} \sigma_x^k z \, dA = B_x \epsilon_{x_{\text{TOTAL}}}^0 + D_x \kappa_{x_{\text{TOTAL}}} - M_{x_T} - M_{x_H} \qquad (4.52)$$

$$Q_x = \sum_{k=1}^{N} \int_{h_{k-1}}^{h_k} \tau_{xz}^{(b)} \, dA \qquad (4.53)$$

The thermal and Hygrothermal * stress resultants are given by,

$$N_{xt} = \sum_{k=1}^{N} \int_{h_{k-1}}^{h_k} E_x^{(k)} \alpha_x^{(k)} \Delta T \, dA, \quad N_{x_H} = \sum_{k=1}^{N} \int_{h_{k-1}}^{h_k} E_x^{(k)} \beta_x^{(k)} \Delta m \, dA \qquad (4.54)$$

$$M_{x_T} = \sum_{k=1}^{N} \int_{k_{k-1}}^{h_k} E_x^{(k)} \alpha_x^{(k)} \Delta T z \, dA, \quad M_{x_H} = \sum_{k=1}^{N} \int_{h_{k-1}}^{h_k} E_x^{(k)} \beta_x^{(k)} \Delta m z \, dA \quad (4.55)$$

In addition to the stress resultants the equations of motion are formed by summing faces along the coordinate directions and equating these expressions,

$$N_{x,x} + p_x(x) = 0$$
$$Q_{x,x} + q(x) = 0 \qquad \text{Equilibrium} \qquad (4.56)$$
$$M_{x,x} - Q_x + m'(x) = 0$$

The equilibrium equations can be rewritten in terms of the displacements

* NOTE: These stress resultants can be due to geometrical constraints (supports) or chemical (bonding of laminae).

resulting in the following set of governing equations,

$$A_x u_{0,xx} - B_x w_{,xxx} + p_x - N_{xT,x} - N_{xH,x} = 0$$

$$B_x u_{0,xxx} - D_x w_{,xxxx} + q + m_x - M_{xT,xx} - M_{xH,xx} = 0 \qquad (4.57)$$

In addition to the governing equations above appropriate boundary conditions must be introduced for completeness. These will be considered by following several examples.

Returning to the basic governing equations for beam bending, we note that the two governing equations in terms of the displacements u_0 and w can be uncoupled in terms of u_0 and w resulting in the following equations,

$$w_{,xxx} = \frac{1}{D_x}[q + m', x - M_{xT,xx} - M_{xT,xx} - M_{xH,xx}] \qquad (4.58)\text{a}$$

$$- \frac{1}{B_x}[p_{x,x} - N_{xT,xx} - N_{xH,xx}]$$

$$u_{0,xxx} = \frac{1}{B_x}[q + m', x - M_{xT,xx} - M_{xH,xx}] \qquad (4.58)\text{b}$$

$$- \frac{1}{A_x}[p_{x,x} + N_{xT,xx} + N_{xH,xx}]$$

In the above equations, the "reduced" stiffnesses due to a non-zero

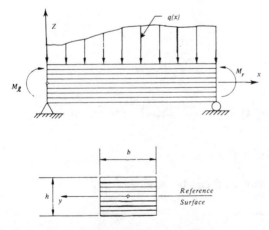

Figure 4.5. Simply Supported Laminated Beam.

B matrix, i.e., asymmetry about the midsurface, are given by

$$\bar{A}_x = \frac{A_x D_x - B_x^2}{D_x}$$

(4.59)

$$\bar{B}_x = \frac{A_x D_x - B_x^2}{B_x}$$

(4.60)

$$\bar{D}_x = \frac{A_x D_x - B_x^2}{A_x}$$

(4.61)

Consider now the transverse loading of a simply supported beam as shown above.

In the absence of in-plane, thermal, moisture, and moment distributed loads the governing equation for the deflection w of a simply supported beam under the action of a transverse load distribution becomes, as described in the previous section,

$$w_{,xxxx} = (q_x/\bar{D}_x)$$

(4.62)a

Similarly, $u_{0,xxx} = q/\bar{B}$

(4.63)b

This equation is identical to that derived from first principles in Reference 1, for bending of laminated plate strips. Inherent in this formulation are the following assumptions in regard to the beam deformations, that is,

$v = 0$

u and $w \neq$ functions of y (width)

and,

$\epsilon_x = u_{0,x} - zw_{,xx}$

$\epsilon_y \equiv 0$

$\gamma_{xy} \equiv 0$

The fourth order governing equation can then be integrated directly to yield the deflection

$$w(x) = q(x) + C_1 x^3 + C_2 x^2 + C_3 x + C_4$$

and the in-plane displacement field can be deduced from,

$$u_0(x) = w_{,x} \frac{1}{B_x} + C_5 x^2 + C_6 x + C_7$$

Introducing the following boundary conditions, (a simply supported situation)

$$w(0) = 0 \qquad N_x(0) = 0$$
$$w(L) = 0 \qquad N_x(L) = 0$$
$$M_x(0) = 0 \qquad u_0(0) = 0$$
$$M_x(L) = 0$$

The solution for $w(x)$ and $u_0(x)$ can then be written as,

$$w(x) = \frac{1}{D_x}\left\{ W(x) + L^2 \cdot W_{,xx}(0)\left[\frac{1}{6}\left(\frac{x}{L}\right)^3 - \frac{1}{2}\left(\frac{x}{L}\right)^2 + \frac{1}{3}\left(\frac{x}{L}\right)\right] \right.$$

$$\left. + L^2 \cdot W_{,xx}(L)\left[\frac{1}{6}\left(\frac{x}{L}\right) - \frac{1}{6}\left(\frac{x}{L}\right)^3\right] + W(0)\left[\frac{x}{L} - 1\right] - W(L)\left(\frac{x}{L}\right) \right\}$$

$$(4.63)$$

$$u_0(x) = \frac{1}{B_x}\left\{ W_{,x}(x) + LW_{,xx}(0)\left[\frac{1}{2}\left(\frac{x}{L}\right)^2 - \left(\frac{x}{L}\right)\right] \right.$$

$$\left. - LW_{,xx}(L)\left[\frac{1}{2}\left(\frac{x}{L}\right)\right] - W_{,x}(0) \right\} \qquad (4.64)$$

The stress components can similarly be written as,

$$\sigma_x^{(k)} = E_x^{(k)}\left[\frac{1}{B_x} - \frac{z}{D_x}\right]\left[W_{,xx}(x) + W_{,xx}(0)\left(\frac{x}{L} - 1\right) - W_{,xx}(L)\left(\frac{x}{L}\right)\right]$$

$$(4.65)$$

$$\tau_{xz}^{(k)} = \frac{1}{t}\left[W_{,xxx} + \frac{1}{L}(W_{,xx}(0) - W_{,xx}(L)) \cdot \int_{h_1}^{z} E_x^{(k)}\left[\frac{1}{B_x} - \frac{z}{D_x}\right]dA \right] \quad (4.66)$$

where

$$W(x) = \iiiint q(x)\,dx\,dx\,dx\,dx$$

For the special case of a uniform load,

$$w(x) = \frac{A_x q_0 L^4}{24(A_x D_x - B_x^2)}\left[\left(\frac{x}{L}\right)^4 - 2\left(\frac{x}{L}\right)^3 + \frac{x}{L}\right] \qquad (4.67)$$

Table 4.1

ply sequences	Displacement components	Loading Types					
		E-glass	Graphite	E-glass	Graphite	E-glass	Graphite
$[0|\pm45|90]_s$	$U_{L/4}$	0	0	0	0	$7.7\times10^{-7}P_0L^2$	$9.2\times10^{-7}P_0L^2$
	$U_{L/2}$	0	0	0	0	$4.4\times10^{-7}P_0L^2$	$2.4\times10^{-7}P_0L^2$
	$W_{L/4}$	$4.04\times10^{-5}q_0L^4$	$1.72\times10^{-5}q_0L^4$	$1.93\times10^{-5}\mathbf{K}L^5$	$0.82\times10^{-5}\mathbf{K}L^5$	0	0
	$W_{L/2}$	$5.67\times10^{-5}q_0L^4$	$2.41\times10^{-5}q_0L^4$	$2.84\times10^{-5}\mathbf{K}L^5$	$1.21\times10^{-5}\mathbf{K}L^5$	0	0
$[0|90|0|90]_s$	$U_{L/4}$	0	0	0	0	$5.7\times10^{-7}P_0L^2$	$2.5\times10^{-7}P_0L^2$
	$U_{L/2}$	0	0	0	0	$3.2\times10^{-7}P_0L^2$	$1.4\times10^{-7}P_0L^2$
	$W_{L/4}$	$3.72\times10^{-5}q_0L^4$	$1.51\times10^{-5}q_0L^4$	$1.78\times10^{-5}\mathbf{K}L^5$	$0.72\times10^{-5}\mathbf{K}L^5$	0	0
	$W_{L/2}$	$5.22\times10^{-5}g_0L^4$	$2.12\times10^{-5}q_0L^4$	$2.61\times10^{-5}\mathbf{K}L^5$	$1.06\times10^{-5}\mathbf{K}L^5$	0	0

ply sequences	Displacement components	Loading Types					
		Thermal				Hygrometric $\Delta m = 1\%$	
		$T = 70\left(1.25 + 0.5\dfrac{z}{h}\right)$		$T = \Delta T$			
		E = glass	Graphite	E = glass	Graphite	E = glass	Graphite
$[0|\pm 45|90]_s$	$U_{L/4}$	$-42.7 \times 10^{-5} L$	$-8.42 \times 10^{-5} L$	$-0.49 \times 10^{-5} \Delta TL$	$-0.1 \times 10^{-5} \Delta TL$	$-12.3 \times 10^{-5} \Delta mL$	$-5.65 \times 10^{-5} \Delta mL$
	$U_{L/2}$	$-85.4 \times 10^{-5} L$	$-16.84 \times 10^{-5} L$	$-0.98 \times 10^{-5} \Delta TL$	$-0.2 \times 10^{-5} \Delta TL$	$-24.5 \times 10^{-5} \Delta mL$	$-11.3 \times 10^{-5} \Delta mL$
	$W_{L/4}$	$-56.3 \times 10^{-5} L^2$	$-5.38 \times 10^{-5} L^2$	0	0	0	0
	$W_{L/2}$	$-75 \times 10^{-5} L^2$	$-7.18 \times 10^{-5} L^2$	0	0	0	0
$[0|90|0|90]_s$	$U_{L/4}$	–	–	–	–	–	–
	$U_{L/2}$	–	–	–	–	–	–
	$W_{L/4}$	–	–	–	–	–	–
	$W_{L/2}$	–	–	–	–	–	–

$$u_0(x) = \frac{B_x q_0 L^3}{24(A_x D_x - B_x^2)} \left[4\left(\frac{x}{L}\right)^3 - 6\left(\frac{x}{L}\right)^2 \right] \tag{4.68}$$

The corresponding stresses are,

$$\sigma_x^{(k)} = \frac{q_0 L^2}{2(A_x D_x - B_x^2)} \left[\left(\frac{x}{L}\right)^2 - \left(\frac{x}{L}\right) \right] E_x^{(k)} [B_x - zA_x] \tag{4.69}$$

$$\tau_{xz}^{(k)} = \frac{q_0 L}{2t(A_x D_x - B_x^2)} \left[2\left(\frac{x}{L}\right) - 1 \right] \int_{h_1}^{z} E_x^{(k)} (B_x - zA_x) dA \tag{4.70}$$

Table 4.1 shows the in-plane and transverse deflections occurring at the points $(L/4)$ and $(L/2)$, are given for the following angle ply graphite composite $[0/\pm 45/90]_s$ and compared with a similar glass system for a number of loading types.
Note: The Glass and graphite examples are for comparable volume fractions of reinforcing fibers.

4.4. Axial Loading for Simply Supported Beams

For this case the beam is considered subjected to the distributed action of a system of axial loads accompanied by concentrated end forces along the direction of the reference axes as shown below,
Considering this case of loading, the governing equations can be reduced to the following sets of equations,

$$w_{,xxxx} = p_{x,x}(x)/\overline{B}_x \tag{4.71}$$

and

$$u_{0,xxx} = p_{x,x}(x)/\overline{A}_x \tag{4.72}$$

The above sets of equations can be integrated directly to yield

$$w(x) = \frac{1}{\overline{B}_x} \iiint p_x(x) dx dx dx + C_1 x^3 + C_2 x^2 + C_3 x + C_4 \tag{4.73}$$

and

$$u_0(x)z - \frac{1}{\overline{A}_x} \iint p_x(x) dx dx + C_5 x^2 + C_6 x + C_7 \tag{4.74}$$

In order to determine a unique solution, the constants C_1 through C_7 must be evaluated from the appropriate boundary conditions. The pre-

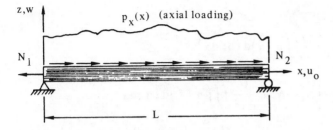

Figure 4.6. Axial Loading of Simply Supported Beam.

scribed boundary conditions for a simply supported beam are,

$$w(0) = w(L) = M_x(0) = M_x(L) = 0$$

$$N_x(0) = N_1, \ N_x(L) = N_2, \ u_0(0) = 0 \qquad (4.75)$$

It should be pointed out that the concentrated end loads and distributed axial loading are not independent but related through the following equilibrium relation

$$N_1 = N_2 + \int_0^L p_x(x)\,\mathrm{d}x \qquad (4.76)$$

The solution for the lateral deflection and the in-plane displacement field are then given by,

$$w(x) = \frac{1}{\bar{B}_x}\left[\iiint p_x(x)\mathrm{d}x\mathrm{d}x\mathrm{d}x + \iiint p_x(0)\mathrm{d}x\mathrm{d}x\mathrm{d}x\right]$$

$$+ \frac{x^3}{6L\bar{B}_x}\left[N_2 - N_1 + \frac{\mathrm{d}^2}{\mathrm{d}x^2}\left(\iiiint p_x(L)\mathrm{d}x\mathrm{d}x\mathrm{d}x - \iiint p_x(0)_x\mathrm{d}x\mathrm{d}x\right)\right]$$

$$+ \frac{x^2}{2\bar{B}_x}\left[N_1 + \frac{\mathrm{d}^2}{\mathrm{d}x^2}\left(\iiiint p_x(0)\mathrm{d}x\mathrm{d}x\mathrm{d}x\right)\right]$$

$$+ \frac{x}{L\bar{B}_x}\left[\iiiint p_x(L)\mathrm{d}x\mathrm{d}x\mathrm{d}x - \iiiint p_x(0)\mathrm{d}x\mathrm{d}x\mathrm{d}x\right.$$

$$\left. - \frac{L^2}{6}\left(N_2 - 2N_1 + \frac{\mathrm{d}^2}{\mathrm{d}x^2}\iiiint p_x(L)\mathrm{d}x\mathrm{d}x\mathrm{d}x + 2\frac{\mathrm{d}^2}{\mathrm{d}x}\iiiint p_x(0)\mathrm{d}x\mathrm{d}x\mathrm{d}x\right)\right]$$

$$(4.77)$$

138

and

$$u_0(x) = \frac{1}{A_x}\left[-\frac{d}{dx}\iiint p_x(x)dxdxdx \right]$$

$$+ \frac{x^2}{2LA_x}\left[N_2 - N_1 + \frac{d^2}{dx^2}\iiint p_x(L)dxdxdx \right.$$

$$\left. + \frac{d^2}{dx^2}\iiint p_x(0)dxdxdx \right] + \frac{x}{A_x}\left[N_1 + \frac{d^2}{dx^2}\iiint p_x(0)dxdxdx \right]$$

$$+ \frac{1}{A_x}\frac{d}{dx}\iiint p_x(0)dxdxdx \tag{4.78}$$

The corresponding stress field is given by,

$$\sigma_x^{(k)} = E_x^{(k)}\left[\frac{1}{A_x} - \frac{z}{B_x} \right]\left[N_1 + \frac{d^2}{dx^2}\iiint p_x(0)dxdxdx \right.$$

$$\left. - \frac{d^2}{dx^2}\iiint p_x(x)dxdxdx \right] \tag{4.79}$$

$$\tau_{xz}^{(k)} = 0 \tag{4.80}$$

For the special case of a uniform axial load as shown in Figure 4.7 below, the deflection, displacement and stress distribution are given by equations (4.81–4.84). Data is shown in Table 4.1 for the case of a graphite and glass epoxy composite

$$u_0(x) = \frac{D_x p_0 L^2}{A_x D_x - B_x^2}\left[-\frac{1}{2}\left(\frac{x}{L}\right)^2 + \frac{x}{L} \right] \tag{4.81}$$

$$w(x) = \frac{B_x p_0 L^3}{6(A_x D_x - B_x^2)}\left[-\left(\frac{x}{L}\right)^3 + 3\left(\frac{x}{L}\right)^2 - 2\left(\frac{x}{L}\right) \right] \tag{4.82}$$

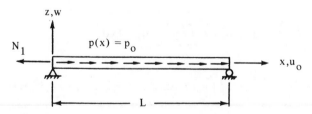

Figure 4.7.

$$\sigma_x^{(k)} = \frac{p_0 L}{A_x D_x - B_x^2} \left(1 - \frac{x}{L}\right) E_x^{(k)} (D_x - zB_x) \tag{4.83}$$

$$\tau_{xz}^{(k)} = 0 \tag{4.84}$$

4.5. Eigenvalue Problems of Beams of Composite Materials: Natural Vibrations and Elastic Stability

To this point the problems studied in this Chapter have concentrated on finding the maximum deflection in composite material beams to insure that they are not too large for a deflection limited or stiffness critical structure, and in determining the maximum stresses in the beam structure for those which are strength critical. However, there are two other ways in which a structure can become damaged or useless; one is through a dynamic response to time dependent loads, resulting again in too large a deflection or too high stresses, and the other is through the occurrence of an elastic instability (buckling).

In the former, dynamic loading on a structure can vary from a re-occurring cyclic loading of the same repeated magnitude, such as a structure supporting an unbalanced motor that is turning at 100 revolutions per minute, for example, to the other extreme of a short time intense, non-reoccurring load, termed shock or impact loading, such as a bird striking an aircraft component during flight. A continuous infinity of dynamic loads exist between these extremes of harmonic oscillation and impact.

A whole volume could and should be written on the dynamic response of composite material structures to time dependent loads, but that is beyond the scope of this text. There are a number of texts dealing with dynamic response of isotropic structures. However, one common thread to all dynamic response will be presented – that of the natural frequencies of vibration.

Any continuous structure mathematically has an infinity of natural frequencies and mode shapes. If a structure is oscillated at a frequency that corresponds to a natural frequency, it will respond by a rapidly growing amplitude with time, requiring very little input energy, until such time as the structure becomes overstressed and fails, or until the oscillations become so large that non-linear effects may limit the amplitude to a large but usually unsatisfactory value because then considerable fatigue damage will occur.

Thus, for any structure, analyzed for its integrity, the natural frequencies should be determined in order to compare them with any time dependent loadings to which the structure will be subjected to insure that the frequencies imposed and the natural frequencies differ considerably. Conversely in designing a structure, over and above insuring that the structure will not overdeflect, or become overstressed, care should be taken to avoid resonances (that is, imposed loads having the same frequency as one or more natural frequencies).

The easiest example to illustrate this behavior is that of the composite material beam previously studied, wherein mid-plane symmetry exists ($B_{ij} = 0$), that is, no bending stretching coupling and no transverse shear deformation ($\epsilon_{xz} = 0$). In this case the governing equations are given by (3.35), as shown below

$$bD_{11}\frac{d^4w}{dx^4} = q(x) \tag{4.86}$$

In (4.86), it is seen that the imposed load is written in force per unit length. If D'Alembert's Principle were used then one could add a term to (4.86) equal to the product of mass and acceleration per unit length. In that case (4.86) becomes

$$bD_{11}\frac{\partial^4w(x,\,t)}{\partial x^4} = q(x,\,t) - \frac{\rho A\partial^2w(x,\,t)}{\partial t^2} \tag{4.87}$$

where, here w and q both become functions of time as well as space, and derivatives therefore become partial derivatives, ρ is the *mass* density of the beam material, and A is the cross-sectional area. In the above $q(x,\,t)$ now is the spatially varying time dependent forcing function causing the dynamic response, and could be anything from a harmonic oscillation to an intense one time impact.

However natural frequencies for the beam occur as functions of the material properties and the geometry and hence are not affected by the forcing functions, hence for their study let $q(x,\,t)$ be zero. Thus, (4.87) becomes:

$$bD_{11}\frac{\partial^4w}{\partial x^4} + \rho A\frac{\partial^2w}{\partial t^2} = 0 \tag{4.88}$$

Assume that the composite beam, if it is simply supported at each end, vibrates with the same deflection function called a vibration mode shape as a simply supported beam of an isotropic material (which in this example is also fact):

$$w(x,\,t) = \sum_{n=1}^{\infty} A_n \sin\frac{n\pi x}{L}\cos \omega_n t \tag{4.89}$$

here A_n is the amplitude, and ω_n is called the natural frequency in radians per unit time for the nth vibrational mode. Note that in this case there is one natural frequency for each natural mode shape, for $n = 1, 2, 3, \ldots$

Substituting (4.89) into (4.88) results in the following equation

$$\sum_{n=1}^{\infty} A_n \left[\frac{n^4 \pi^4}{L^4} b D_{11} - \omega_n^2 \rho A \right] \sin \frac{n\pi x}{L} \cos \omega_n t = 0$$

For this to be true, then for each value of n,

$$\omega_n = \frac{n^2 \pi^2}{L^2} \sqrt{\frac{b D_{11}}{\rho A}} \tag{4.90}$$

Thus, for each n there is a different natural frequency, and a different mode shape, as shown in Figure 4.9.

For each value of n, the natural frequency is important because if the beam were oscillating at that particular frequency, a resonance would occur, and the amplitude would grow until some form of structural failure would occur.

The lowest natural frequency, corresponding to $n = 1$, is termed the

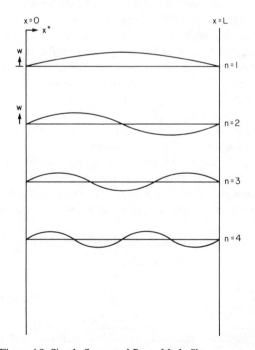

Figure 4.9. Simply Supported Beam Mode Shapes.

fundamental frequency. It should be noted that n could go from a value of one to infinity. However the governing differential equation (4.88) is only applicable over a portion of this range. For an isotropic single layer beam, the equation breaks down when the half wave length becomes close to the height, h, of the beam, because then transverse shear deformation effects (ϵ_{xz} and $\epsilon_{yz} \neq 0$) become important. In composite material beams transverse shear deformation effects can be important even for the fundamental natural frequency, but to include transverse shear deformation effects involves considerable analytical complications, which will be dealt with later in this Chapter. Equation (4.90) is useful, however, for preliminary design.

Incidentally, one sees that for each value of ω_n there is one function $w(x, t)$, hence, the word characteristic function, or (half German) eigenfunction. The natural frequency is then the eigenvalue for that eigenfunction. Note that with this theory (the eigenvalue problem) the amplitude A_n cannot be determined, and the governing differential equation (4.88) is homogeneous.

In addition to looking at maximum deflections, maximum stresses and natural frequencies when analyzing a structure, one must investigate under what loading conditions an instability can occur, which is also generically referred to as buckling.

Looking at the governing equations for a composite beam with mid-plane symmetry ($B_{ij} = 0$) given by (4.13) through (4.17), it is seen that there are two equations dealing with in-plane loads P and in-plane deflections u_0, namely (4.13) and (4.15); (4.18) being the solution result. Likewise the remaining three equations deal only with the lateral load $q(x)$ and the lateral deflection w that results, with (4.27) being the result for the particular lateral loading $q(x) = -q_0$. So the in-plane action is entirely independent of the lateral action.

Yet it is well known and often observed that in-plane loads through buckling do cause lateral deflections, which are usually disastrous. The answer to the paradox is that entirely through this Chapter up to this point we have used a *linear* elasticity theory, and the physical event of buckling is a non-linear theory problem. For brevity, the development of the non-linear theory will not be included herein because it is included in so many other texts [2,3].

The result of including the terms to predict the advent or inception of buckling for the beam is:

$$bD_{11} \frac{d^4 w}{dx^4} = q(x) + P \frac{d^2 w}{dx^2} \tag{4.91}$$

where clearly there is a coupling between the in-plane load and the lateral deflection.

Some items should be noted. The buckling loads like the natural frequencies in vibration are independent of the lateral loads, which will be disregarded in what follows. However, in actual structural analysis, the effect of lateral loads, along with the in-plane loads could cause overstressing and failure before the in-plane buckling load is reached. However, the buckling load is independent of the lateral load, as are the natural frequencies. Incidentally, common sense dictates that if one is designing a structure to withstand compressive loads, with the possibility of buckling being the failure mode, one had better design the structure to be mid-plane symmetric, so that $B_{ij} = 0$, otherwise the bending stretching coupling would likely cause overstressing before the buckling load is reached, as well as reduce the buckling load itself, see (4.95).

Looking at the buckling of a beam only, (4.91) can be written as

$$bD_{11} \frac{d^4 w}{dx^4} = P \frac{d^2 w}{dx^2} \tag{4.92}$$

One can assume that the buckling mode shapes $w_n(x)$ for the composite beam which is simple supported at each end, is the same as that for the buckling modes of an isotropic beam with the same boundary conditions which is discussed in many texts, namely,

$$w(x) = \sum_{n=1}^{\infty} A_n \sin \frac{n\pi x}{L} \tag{4.93}$$

Substituting (4.93) into (4.92) yields

$$\sum_{n=1}^{\infty} A_n \left[bD_{11} \frac{n^4 \pi^4}{L^4} + P \frac{n^2 \pi^2}{L^2} \right] \sin \frac{n\pi x}{L} = 0$$

and for this to be true for all n, the critical values of P, called P_{cr} are

$$P_{cr} = - \frac{n^2 \pi^2}{L^2} bD_{11} \tag{4.94}$$

Several things are clear: (4.92) is a homogeneous equation, one cannot obtain the magnitude A_n of the buckling mode, and unlike in vibrations were many values of n are important, here only the one giving the lowest value of P_{cr} is of importance, because after buckling the column is usually permanently deformed or fails, and finally, P_{cr} is a negative value which means a compressive inplane load.

Thus, it is seen that the eigenvalue problems differ from the considerations in previous sections because the event, natural vibration or buckling, occur only at certain values. Hence, the natural frequencies and the buckling load are the eigenvalues; the vibrational mode and the buckling mode are the eigenfunctions.

In analyzing any structure one therefore should determine four things: the maximum deflection, the maximum stresses, the natural frequencies (if there is any dynamic loading to the structure – or nearby to the structure) and the buckling loads (if there are any compressive loads).

For composite beams with mid-plane symmetry, and no other coupling terms (16 terms equal to zero), one can use the vibrational modes or buckling modes respectively for isotropic beams in the x-direction that have the appropriate boundary condition.

All combinations of beam vibrational mode shapes applicable for use here have been developed by Warburton, and all derivatives and integrals of those functions catalogued conveniently by Young and Felgar for easy use, as referenced in Chapter 3.

Similar expressions for buckling modes are in general available also in many texts, so likewise they can be used in (4.92).

It must again be stated that the natural frequencies and buckling loads calculated in this section do not include transverse shear effects, and are therefore only approximate – but they are useful for preliminary design, because of their relative simplicity. If the transverse shear deformation were included, see the next Section, both natural frequencies and buckling loads would be lower than those calculated in this Section – so the buckling loads calculated, neglecting transverse shear deformation, are not conservative.

It has been shown that in the equation for natural frequencies and buckling loads, (that is, the eigenvalues), that if the laminate is unsymmetric with respect to the middle surface (that is, $B_{11} \neq 0$), then the D_{11} terms in those equations can be replaced by

$$\frac{A_{11} D_{11} - B_{11}^2}{A_{11}} \tag{4.95}$$

Equation (4.95) has been termed the "reduced" or "apparent" flexural stiffness.

Likewise it has been shown that the simple expressions for approximate natural frequencies and buckling loads, which do not include transverse shear deformation effects, can be modified to initiate such effects for the columns discussed previously. The result is

$$\left\{ \begin{matrix} \omega_n \\ P_{cr} \end{matrix} \right\}_{\text{with tsd}} = \frac{\left. \begin{matrix} \omega_n \\ P_{cr} \end{matrix} \right\} \text{ previously determined}}{1 + n \left(\dfrac{\pi}{2L} \right)^2 \dfrac{D_{11}}{D_{44}}} \tag{4.96}$$

4.6. Thermoelastic Effects on Beams of Composite Materials [Moderate such that the mechanical properties remain virtually unchanged for the temperatures considered.]

In the preceding section the effect of mechanical loads of different type acting upon laminated beams was developed. The theory and resultant expressions as derived are valid within the framework of the classical theory of beam bending. In the present section the effect of thermally induced stresses and deformation are examined. Since the problems dealt with represent linear elastic type solutions the principal of superposition of solutions is valid and can be used to calculate the appropriate quantities of interest.

The governing equations for the deflection and mid-plane displacement, a steady state temperature field is given by,

$$w_{,xxxx} = -M_{xT,xx}/\overline{D}_x + N_{xT,xx}/\overline{B}_x \tag{4.97}$$

and

$$u_{0,xxx} = -M_{xT,xx}/\overline{B}_x + N_{xT,xx}/\overline{A}_x \tag{4.98}$$

where M_{xT} and N_{xT} are the thermal stress couple and the thermal stress resultant of the beam. The above equations can be integrated directly to yield,

$$w(x) = -M_T/\overline{D}_x + N_T/\overline{B}_x + C_1 x^3 + C_2 x^2 + C_3 x + C_4 \tag{4.99}$$

and

$$u_0(x) = -M_{T,x}/\overline{B}_x + N_{T,x}/\overline{A}_x + C_5 x^2 + C_6 x + C_7 \tag{4.100}$$

In the above equations,

$$M_T(x) = \iint M_{xT}(x)\mathrm{d}x\mathrm{d}x \tag{4.101}$$

$$N_T(x) = \iint N_{xT}(x)\mathrm{d}x\mathrm{d}x \tag{4.102}$$

The integration constants C_1 thru C_7 are determined from the beam boundary conditions. For a simple supported beam of the type shown in Figure 4.5 the boundary or constraint conditions can be written as

$$w(0) = w(L) = M_x(0) = M_x(L) = 0$$

$$N_x(0) = N_x(L) = u_0(0) = 0$$

The constants can thus be evaluated and are given by the following results,

$$C_1 = C_2 = C_5 = C_6 = 0$$

$$C_3 = [M_T(0) - M_T(L)]/L\bar{B}_x + [m_T(L) - m_T(0)]/L\bar{D}_x$$

$$C_4 = -M_T(0)/\bar{B}_x + m_T(0)/\bar{D}_x$$

$$C_7 = -M_{T,x}(0)/\bar{A}_x + m_{T,x}(0)/\bar{B}_x$$

Substituting the constants back into the integrated governing equations, the following deflection equation can be determined,

$$w(x) = \frac{1}{\bar{B}_x}\left[M_T(x) - M_T(L)\left(\frac{x}{L}\right) - M_T(0)\left(1 - \frac{x}{L}\right)\right]$$

$$- \frac{1}{\bar{D}_x}\left[m_T(x) - m_T(L)\left(\frac{x}{L}\right) - m_T(0)\left(1 - \frac{x}{L}\right)\right] \qquad (4.103)$$

and

$$u_0(x) = \frac{1}{\bar{A}_x}[M_{T,x}(x) - M_{T,x}(0)] - \frac{1}{\bar{B}_x}[m_{T,x}(x) - m_{T,x}(0)] \qquad (4.104)$$

Some numerical results for graphite and glass epoxy composite beams subjected to a thermal field are incorporated in Table 4.1. The corresponding difference in deflections should be noted and compared with the case of the simply supported beam subjected to mechanical loads only.

The induced stresses due to the imposed strain field can be calculated from the relation

$$\sigma_x^{(k)} = E_x^{(k)}\epsilon_x^{(k)}, \quad \tau_{xz}^{(k)} = \frac{1}{t}\int_{h_0}^z \sigma_{x,x}^{(k)}dA. \qquad (4.105)$$

In the case of a hygrothermal environment imposed upon a non-metal matrix composite the same calculation procedure can be used with simple interchange of moisture diffusion for the thermal effect noted. Once again caution must be noted in this simplifying analytical process due to the potential effect of using variations in mechanical properties as effected by moisture and temperature.

4.7. Problems

4.1. The governing differential equation for a beam composed of a composite material in which $B_{ij} = 0$ and $D_{16} = D_{26} = 0$ is

$$bD_{11}\frac{d^4w}{dx^4} = q(x)$$

where b is the beam width, D_{11} is the flexural stiffness in the x direction, and $q(x)$ is the lateral load per unit length. If the beam is of length L, and the beam is simply supported at $x = 0$, clamped at $x = L$, and the lateral load is a constant $q(x) = q_0$, determine explicitly the expression $w(x)$.

4.2. Given a beam made of T300/5208 graphite epoxy with the following mechanical properties at 70°F.

$$E_1 = 14.5 \times 10^4 \text{ MPa} = 21 \times 10^6 \text{ psi}$$

$$E_2 = 1.17 \times 10^4 \text{ MPa} = 1.76 \times 10^6 \text{ psi}$$

$$G_{12} = 0.45 \times 10^4 \text{ MPa} = 0.65 \times 10^6 \text{ psi}$$

$$\nu_{12} = 0.21, \ \nu_{21} = 0.017,$$

$$\rho = 0.06 \text{ lbs/in}^3, \text{ the weight density}$$

$$g = 386 \text{ in/sec}^2$$

assume $\sigma_{\text{allowable}} = 100{,}000$ psi

Consider the beam to be made of an all 0° layup, of 30 laminae, each 0.006″ thick, such that the total beam thickness is $h = 0.18″$. The beam is one inch wide ($b = 1″$) and twelve inches long ($L = 12″$).

(a) If the beam is simply supported at each end, and subjected to a uniform lateral load of 10 lbs./in. of length, what is the maximum deflection?

(b) What is its maximum stress?

(c) What is its fundamental natural frequency in bending?

(d) If the beam were subjected to a compressive end load what would the critical buckling load be?

4.3. Consider a beam of length 20″, width 1″ of the material of Problem 4.2. If the beam is simply supported at each end, what minimum beam thickness, h, is necessary to insure that the beam is not overstressed when it is subjected to a uniform lateral load of $q_0 = 20$ lbs./in.?

4.4 For a beam of the material of Problem 4.2, with $L = 15″$, $b = 1″$ and $h = 0.020″$, what is the fundamental natural frequency in cycles per second (Hz) if the beam is simply supported at each end? Neglect transverse shear deformation.

4.5 For the beam of Problem 4.4 above, what is the critical buckling load, P_{cr}, neglecting transverse shear deformation?

5. Composite Material Shells

5.1. Introduction

A shell is a thin walled body, just as a beam or plate is, whose middle surface is curved in at least one direction. For instance a cylindrical shell and a conical shell have only one direction in which the middle surface is curved. On the other hand in a spherical shell there is curvature in both directions. Such mundane shells as a front fender of a car or an egg shell are example of double curvature in shells.

Shell theory is greatly complicated, compared to beam and plate theory because of this curvature. The treatment of shell theory in its proper detail is the subject of a graduate level full semester or full year course, and hence, well beyond the scope of this book. Even to derive the governing differential equations for a shell of general curvature from first principles require several lectures in topology.

Then, to complicate the shell theory with all of the material complexities associated with laminated composite materials makes shell theory of composite materials very complicated, and a great challenge.

5.2. History of composite Material Shells

A rather complete historical treatise on the analysis of composite material shells, both utilizing classical shell theory and shells including transverse shear deformation through 1974 has been compiled by Vinson and Chou [1]. Therefore, this section will treat only the advances made since that time in the static analysis of composite shells.

In 1975, Wilkins and Love [2] examined the combined compression and shear buckling behavior of laminated composite cylindrical shells characteristic of fuselage structures. Boron-epoxy and graphite epoxy shells of both $[\pm 45°]_s$ and $[0°, \pm 45°]_s$ laminations were tested. Specimen sizes were 15″ diameter and 15″ length, with wall thickness of $0.0212″ - 0.0336″$. Compression-shear interaction curves were obtained for all of the above. Compared to classical buckling theory, the actual compression buckling values were 65% of the theoretical value. The disparity was attributed to imperfections. Good agreement between theory and experiment were realized for shear buckling. It was observed that the compression-shear interaction was essentially linear.

In 1976, Waltz and Vinson [3] presented methods of analysis for the determination of interlaminar stresses in laminated cylindrical shells of composite materials. This will be discussed later in the chapter.

Also in 1976, El Naschie [4] investigated the large deflection behavior of composite material shells in determining the lower limit of the asymmetric buckling load.

In 1977, Ecord [5] wrote on a very practical composite shell structure, namely pressure vessels for the space shuttle orbiter. Here, a Kevlar 49 overwrap is used over a titanium and Inconel spherical pressure vessel structure. The Kevlar overwrap was designed to retain the internal pressure without the metallic liner. By using the Kevlar 49 overwrap 400 lbs. are saved in the orbiter; 1750 lbs. without composites, 1350lbs. with the Kevlar 49 overwrap.

In 1978, many notable papers were published concerning composite material shells.

Johnson, Reck, and Davis [6] published a paper dealing with the design, fabrication and testing of a 10 foot long, 10 foot diameter ring stiffened corrugated graphite-epoxy cylindrical shell, typical of a large space structure, capable of resisting buckling. The results of the project established the feasibility of efficiently utilizing composites in structural shell applications. Compared with an aluminum design for the same application, the use of composites resulted in a 23% weight reduction.

Fujczak [7] studied the torsional fatigue behavior of graphite-epoxy cylindrical shells. New important information resulted from that study.

In 1978, Booton [8] investigated the buckling of imperfect composite material cylinders under combined loadings, both theoretically and experimentally. The combined loadings involved axial compression, external pressure and torsion. Donnell-Mushtari theory was used. Imperfections were more critical in axial compression than in external pressure or torsion, as is to be expected.

Raju, Chandra and Rao [9] studied the determination of transient temperatures in laminated composite conical shells caused by aerodynamic heating.

In 1978, Varadan [10] studied the snap-through buckling of composite shallow spherical shells. He calculated the critical buckling external pressure as a function of both the shell geometry and material properties. Also, Rhodes and Marshall [11] studied the asymmetric buckling of laterally loaded composite material shells. Montague [12] experimentally studied the behavior of double-skinned composite, circular cylindrical shells subjected to external pressure.

In 1979, Humphrey [13] experimentally investigated hygrothermal effects on composite material pressure vessels, to be used as rocket motor cases. The tests show that Kevlar composites suffer far less degradation than fiberglass.

More recently, Bert [14,15] and his colleagues have been very prolific in the area of shell theory of composite materials. He has to some degree concentrated in the behavior of composites which have different properties in tension and compression – which he terms bi-modulus composites. These are typical of some composites such as fiber reinforced tires, and some biological materials.

Also, Yuceoglu and Updike [16] have studied the stress concentrations in bonded, multi-layer cylindrical shells.

5.3. Analysis of Composite Material Cylindrical Shells Under Axially Symmetric Loads

The simplest of all shell geometries is that of the circular cylindrical shell shown below in Figure 5.1. The positive directions of the displacements u, v, and w are shown, as well as the positive directions of the coordinates x and ζ. The remaining coordinate is the circumferential coordinate θ. The positive value of all stress resultants and stress couples are shown in Figure 5.2 below.

In the classical shell theory discussed here, all of the assumptions used in plate theory of Chapters 2 and 3 are utilized:

$$\sigma_\zeta = \epsilon_\zeta = \epsilon_{x\zeta} = \epsilon_{\theta\zeta} = 0$$

$$u(x, \theta, \zeta) = u_0(x, \theta) + \zeta\beta_x(x, \theta) \tag{5.1}$$

$$v(x, \theta, \zeta) = v_0(x, \theta) + \zeta\beta_\theta(x, \theta)$$

where all of the terms have been explained in the previous chapters.

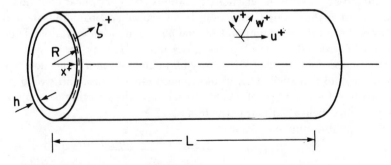

Figure 5.1. Circular Cylindrical Shell.

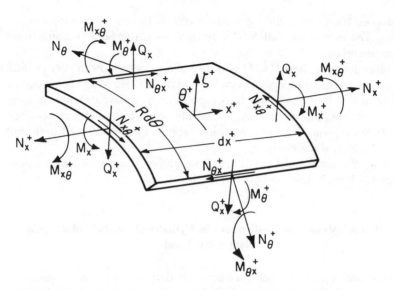

Figure 5.2. Positive Directions of Stress Resultants and Stress Couples.

In addition, there is another assumption known as Love's First Approximation, which is consistent with the neglect of transverse shear deformation:

$$h/R \ll 1. \tag{5.2}$$

It is true that the accurate analysis of shells of composite materials should include transverse shear deformation because of the fact that the moduli of elasticity in the fiber direction is a fiber dependent property, while the transverse shear modulus is a matrix dominated property. However, for preliminary design, one can neglect transverse shear deformation, with the resulting simplifications.

To derive the governing differential equations for cylindrical shells, one should begin with the elasticity equations in a curvilinear coordinate system, just as the elasticity equations in a Cartesian coordinate system in Chapter 3 were the starting point for the plate equation derivation. Then one could proceed to develop the governing equations for a shell of general shape, then specialize it to any particular configuration such as a circular cylindrical shell. This would require much space, and would be too involved for the scope of this text. Therefore, the governing equations are presented below without derivation.

The equilibrium equations are:

$$\frac{\partial N_x}{\partial x} + \frac{1}{R} \frac{\partial N_{x\theta}}{\partial \theta} + q_x = 0 \tag{5.3}$$

$$\frac{\partial N_{x\theta}}{\partial x} + \frac{1}{R}\frac{\partial N_\theta}{\partial \theta} + \frac{Q_\theta}{R} + q_\theta = 0 \qquad (5.4)$$

$$\frac{\partial Q_x}{\partial x} + \frac{1}{R}\frac{\partial Q_\theta}{\partial \theta} - \frac{1}{R}N_\theta + p(x,\theta) = 0 \qquad (5.5)$$

$$\frac{\partial M_x}{\partial x} + \frac{1}{R}\frac{\partial M_{x\theta}}{\partial \theta} - (Q_x - m_x) = 0 \qquad (5.6)$$

$$\frac{\partial M_{x\theta}}{\partial x} = \frac{1}{R}\frac{\partial M_\theta}{\partial \theta} - (Q_\theta - m_\theta) = 0 \qquad (5.7)$$

where

$$q_x = \sigma_{zx}(h/2) - \sigma_{zx}(-h/2) = \tau_{1x} - \tau_{2x}$$

$$q_\theta = \sigma_{z\theta}(h/2) - \sigma_{z\theta}(-h/2) = \tau_{1\theta} - \tau_{2\theta}$$

$$m_x = \frac{h}{2}\left[\sigma_{zx}(h/2) + \sigma_{zx}(-h/2)\right] = \frac{h}{2}\left[\tau_{1x} + \tau_{2x}\right]$$

$$m_\theta = \frac{h}{2}\left[\sigma_{z\theta}(h/2) + \sigma_{z\theta}(-h/2)\right] = \frac{h}{2}\left[\tau_{1\theta} + \tau_{2\theta}\right].$$

These equilibrium equations are independent of the material system. They could be written in terms of ∂s, arc distance where $\partial s = R\partial \theta$. The q_x, q_θ, m_x and m_θ quantities are functions of the surface shear stresses.

For the case of no transverse shear deformation, the relation between the rotations β_x and β_θ and the displacements is given by:

$$\beta_x + \frac{\partial w}{\partial x} = 0 \qquad (5.8)$$

$$\beta_\theta + \frac{1}{R}\frac{\partial w}{\partial \theta} - \frac{v_0}{R} = 0. \qquad (5.9)$$

For simplicity, we will assume that the lamination stacking sequence is such that there is no bending-stretching coupling, (that is $[B] = 0$) and that there are no other coupling terms, that is $(\ )_{16} = (\ )_{26}) = 0$. Also, for the following derivation, the simplest case will be studied – that of a shell of one lamina only. Subsequently, generalizations can be made. In this case, the stress-strain relations and the strain-displacement relations, utilizing the displacement assumptions of (5.1) are:

$$\epsilon_x = \frac{1}{E_x}[\sigma_x - \nu_{x\theta}\sigma_\theta] = \frac{\partial u_0}{\partial x} + \zeta\frac{\partial \beta_x}{\partial x} \qquad (5.10)$$

$$\epsilon_\theta = \frac{1}{E_\theta}[\sigma_\theta - \nu_{\theta x}\sigma_x] = \frac{1}{R}\left[\frac{\partial v_0}{\partial \theta} + w\right] + \frac{\zeta}{R}\frac{\partial \beta_\theta}{\partial \theta} \qquad (5.11)$$

$$\epsilon_{x\theta} = \frac{1}{2G_{x\theta}}\sigma_{x\theta}. \qquad (5.12)$$

The in-plane stiffnesses and the flexural stiffnesses are given as follows:

$$K_x = \frac{E_x h}{(1 - \nu_{x\theta}\nu_{\theta x})}, \qquad K_\theta = \frac{E_\theta h}{(1 - \nu_{x\theta}\nu_{\theta x})}$$

$$D_x = \frac{E_x h^3}{12(1 - \nu_{x\theta}\nu_{\theta x})}, \qquad D_\theta = \frac{E_\theta h^3}{12(1 - \nu_{x\theta}\nu_{\theta x})}. \qquad (5.13)$$

The orthotropic relationship of course holds

$$\frac{\nu_{x\theta}}{E_x} = \frac{\nu_{\theta x}}{E_\theta} \qquad (5.14)$$

We now assume that the loads are axially symmetric hence, $\partial(\ )/\partial\theta = 0$.

Doing the usual integration of (5.10) and (5.11) by multiplying each by dz and integrating from $-h/2$ and $h/2$, and utilizing the definitions of the stress resultants results in:

$$N_x - \nu_{x\theta}N_\theta = E_x h \frac{\partial u_0}{\partial x}, \qquad N_\theta - \nu_{\theta x}N_x = E_\theta h \frac{w}{R}.$$

Rearranging these, the following emerges

$$N_x = K_x\left[\frac{\partial u_0}{\partial x} + \nu_{\theta x}\frac{w}{R}\right] \qquad (5.15)$$

$$N_\theta = K_\theta\left[\nu_{x\theta}\frac{\partial u_0}{\partial x} + \frac{w}{R}\right]. \qquad (5.16)$$

Similarly, multiplying (5.10) and (5.11) by $\zeta d\zeta$ and integrating from $-h/2$ to $h/2$, results in:

$$M_x - \nu_{x\theta}M_\theta = -\frac{E_x h^3}{12}\frac{\partial^2 w}{\partial x^2}$$

$$M_\theta - \nu_{\theta x}M_x = 0.$$

Rearranging these yields the following:

$$M_x = -D_x \frac{\partial^2 w}{\partial x^2} \tag{5.17}$$

$$M_\theta = v_{\theta x} M_x. \tag{5.18}$$

We now further assume for the following derivation that there are no surface shear stresses, hence,

$$q_x = q_\theta = m_x = m_\theta = 0.$$

The above shell equations can be simplified to the following:

$$\frac{dN_x}{dx} = 0 \tag{5.19}$$

$$\frac{dQ_x}{dx} - \frac{N_\theta}{R} + p(x) = 0 \tag{5.20}$$

$$\frac{dM_x}{dx} - Q_x = 0 \tag{5.21}$$

$$\beta_x + \frac{dw}{dx} = 0 \tag{5.22}$$

$$\beta_\theta = 0 \tag{5.23}$$

$$N_x = K_x \left[\frac{du_0}{dx} + \frac{v_{\theta x}}{R} w \right] \tag{5.24}$$

$$N_\theta = K_\theta \left[v_{x\theta} \frac{du_0}{dx} + \frac{w}{R} \right] \tag{5.25}$$

$$M_x = -D_x \frac{d^2 w}{dx^2} \tag{5.26}$$

$$M_\theta = v_{\theta x} M_x \tag{5.27}$$

$$Q_x = -D_x \frac{d^3 w}{dx^3}. \tag{5.28}$$

From (5.19) and (5.24),

$$\frac{d^2 u_0}{dx^2} + \frac{v_{\theta x}}{R} \frac{dw}{dx} = 0. \tag{5.29}$$

From (5.28), (5.25) and (5.20), one obtains:

$$-D_x \frac{d^4 w}{dx^4} - \frac{K_\theta}{R}\left[\nu_{x\theta} \frac{du_0}{dx} + \frac{w}{R} \right] + p(x) = 0. \tag{5.30}$$

From (5.29) and (5.30) the resulting governing equation is:

$$\frac{d^4 w}{dx^4} + \frac{12(1 - \nu_{x\theta}\nu_{\theta x})}{h^2 R^2} \frac{D_\theta}{D_x} w = \frac{1}{D_x}\left[p(x) - \nu_{\theta x}\frac{N_x}{R} \right]. \tag{5.31}$$

Now defining

$$\epsilon^4 \equiv \frac{3(1 - \nu_{x\theta}\nu_{\theta x})}{h^2 R^2} \frac{D_\theta}{D_x}, \tag{5.32}$$

the governing differential equations for the lateral deflection w, and the in-plane displacements u_0 become

$$\frac{d^4 w}{dx^4} + 4\epsilon^4 w = \frac{1}{D_x}\left[p(x) - \frac{\nu_{\theta x}}{R} N_x \right] \tag{5.33}$$

$$\frac{d^2 u_0}{dx^2} + \frac{\nu_{\theta x}}{R} \frac{dw}{dx} = 0. \tag{5.34}$$

The form of (5.33) is desirable since it is uncoupled from the other governing equation (5.34). N_x is a constant (see (5.19)) determined by boundary conditions. In fact, it is seen from (5.33) that the presence of an axial in-plane force is that of an equivalent lateral pressure as far as the lateral displacement w is concerned.

Upon determining w and u_0 from (5.33) and (5.34), the stress resultants and stress couples can be obtained from (5.24) and (5.28). Stresses can then be determined from the following

$$\sigma_x = \frac{N_x}{H} + \frac{M_x \zeta}{h^3/12} \tag{5.35}$$

$$\sigma_\theta = \frac{N_\theta}{h} + \frac{M_\theta \zeta}{h^3/12} \tag{5.36}$$

$$\sigma_{xz} = \frac{3Q_x}{2h}\left[1 + \left(\frac{\zeta}{h/2} \right)^2 \right] \tag{5.37}$$

It is also seen that the governing differential equation (5.33) has the same form as that of a beam on an elastic foundation, that is, by replacing D_x by the flexural stiffness of a beam, EI, and replacing $4\epsilon^4 D_x$ by k, the beam foundation modulus. Thus, the physical intuition of all of the solutions for beams on an elastic foundation can be utilized.

By standard methods, the roots of the fourth order equation (5.33) are obtained to be $\pm\epsilon(1 \pm i)$ where $i = \sqrt{-1}$. The general solution can be written as

$$w(x) = A\, e^{-\epsilon x}\cos \epsilon x + B\, e^{-\epsilon x}\sin \epsilon x + C\, e^{\epsilon x}\cos \epsilon x$$

$$+ E\, e^{\epsilon x}\sin \epsilon x + w_p(x) \tag{5.38}$$

where A, B, C, and E are constants of integration determined by the boundary conditions, and $w_p(x)$ is the particular integral.

The in-plane displacement $u_0(x)$ can be obtained through integrating (5.24) as follows

$$u_0(x) = \frac{N_x x}{K_x} - \frac{v_{\theta x}}{R}\int w\,\mathrm{d}x + F \tag{5.39}$$

where N_x is constant and F is a constant of integration.

It is seen, therefore, that for circular cylindrical shells under axially symmetric loads there are six boundary conditions, three at each end. The natural boundary conditions for this case are:

Either u_x is prescribed or $N_x = 0$

Either $\dfrac{\mathrm{d}w}{\mathrm{d}x}$ is prescribed or $M_x = 0$

Either w is prescribed or $Q_x = 0$.

In the above, the simply supported, clamped and free edges are contained.

In solving (5.33), another form may be utilized, other than that of (5.38), the so-called bending boundary layer solution. In that case, the solution can be written as:

$$w(x) = \frac{M_0}{2\epsilon^2 D_x}e^{-\epsilon x}(\sin \epsilon x - \cos \epsilon x) - \frac{Q_0}{2\epsilon^3 D_x}e^{-\epsilon x}\cos \epsilon x$$

$$+ \frac{M_L}{2\epsilon^2 D_x}e^{-\epsilon(L-x)}\left[\sin \epsilon(L - x) - \cos \epsilon(L - x)\right]$$

$$+ \frac{Q_L}{2\epsilon^3 D_x}e^{-\epsilon(L-x)}\cos \epsilon(L - x) + w_p(x) \tag{5.40}$$

where instead of A, B, C and E of (5.38) being the constants of integration M_0, Q_0, M_L and Q_L are the integration constants.

The particular advantage of using (5.40) rather than (5.38) is easily seen. It can be shown that each term of the homogeneous solution contains trigonometric terms, which oscillate between ± 1, multiplied by an exponential term with a negative exponent, hence an exponential decay. If we say that any term is negligible when $e^{-\epsilon x} \leqslant 0.006$ or $e^{-\epsilon(L-x)} \leqslant 0.006$. This occurs when $\epsilon x = \epsilon(L-x) > 5.15$. If $\nu_{\theta x}\nu_{x\theta}$ were 0.09 then, that condition is met whenever

$$x > 4\sqrt{Rh(D_x/D_\theta)^{1/2}}$$
$$L - x > 4\sqrt{Rh(D_x/D_\theta)^{1/2}} \, . \tag{5.41}$$

This is most important because away from either end of the shell a distance greater than $4\sqrt{Rh(D_x/D_\theta)^{1/2}}$ the entire homogeneous solution of (5.40) goes ostensibly to zero, and is neglected. In the case of continuous loads, $p(x)$, the particular solution is one often leading only to membrane stresses and displacements. Because of this the region $0 \leqslant x \leqslant 4\sqrt{Rh(D_x/D_\theta)^{1/2}}$ and $0 \leqslant L - x \leqslant 4\sqrt{Rh(D_x/D_\theta)^{1/2}}$ is called the "bending boundary layer." It is seen that the length of the bending boundary layer is a function of the flexural stiffness D_x and D_θ.

Computationally, (5.40) is very useful. For a shell longer than the bending boundary layer, that is, $L \geqslant 4\sqrt{Rh(D_x/D_\theta)^{1/2}}$ in the region $0 \leqslant x \leqslant 4\sqrt{Rh(D_x/D_\theta)^{1/2}}$, the terms involving M_L and Q_L are neglected; in the region $0 \leqslant (L-x) \leqslant 4\sqrt{Rh(D_x/D_\theta)^{1/2}}$ the terms involving M_0 and Q_0 can be ignored; and in the region $4\sqrt{Rh(D_x/D_\theta)^{1/2}} \leqslant x \leqslant L - 4\sqrt{(Rh/1)(D_x/D_\theta)^{1/2}}$, the terms involving M_0, Q_0, M_L, Q_L can be ignored. Also, when the shell is longer than the bending boundary layer and when $d^2p(x)/dx^2 = d^3p(x)/dx^3 = 0$, then and only then,

$$M_0 = M_x(0), \quad M_L = M_x(L)$$
$$Q_0 = Q_x(0), \quad Q_L = Q_x(L).$$

Even though the above derivation is for a single-layer orthotropic material, it can be easily extended to that of a laminated composite shell, if all couplings are zero. In that case, the general solution can be written as follows.

5.4. A General Solution for Composite Cylindrical Shells Under Axially Symmetric Loads

$$w(x) = \frac{M_0}{2\epsilon^2 D_{11}} e^{-\epsilon x}(\sin \epsilon x - \cos \epsilon x) - \frac{Q_0}{2\epsilon^2 D_{11}} e^{-\epsilon x} \cos \epsilon x$$

$$+ \frac{M_L}{2\epsilon^2 D_{11}} e^{-\epsilon(L-x)} [\sin \epsilon(L-x) - \cos \epsilon(L-x)]$$

$$+ \frac{Q_L}{2\epsilon^3 D_{11}} e^{-\epsilon(L-x)} \cos \epsilon(L-x) + \frac{1}{4\epsilon^4 D_{11}} \left[p(x) - \frac{\nu_{\theta x}}{R} N_x \right]$$

$$(5.42)$$

where

$$\epsilon^4 = \frac{3(1 - \nu_{\theta x}\nu_{\theta x})}{R^2 h^2} \frac{D_{22}}{D_{11}}$$

$$\frac{dw}{dx} = \frac{M_0}{\epsilon D_{11}} e^{-\epsilon x} \cos \epsilon x + \frac{Q_0}{2\epsilon^2 D_{11}} e^{-\epsilon x}(\sin \epsilon x + \cos \epsilon x)$$

$$- \frac{M_L}{\epsilon D_{11}} e^{-\epsilon(L-x)} \cos \epsilon(L-x)$$

$$+ \frac{Q_L}{2\epsilon^2 D_{11}} e^{-\epsilon(L-x)} [\sin \epsilon(L-x) + \cos \epsilon(L-x)]$$

$$+ \frac{1}{4\epsilon^4 D_{11}} \frac{dp(x)}{dx} \qquad (5.43)$$

$$M_x = -D_{11} \frac{d^2 w}{dx^2} = M_0 e^{-\epsilon x}(\sin \epsilon x + \cos \epsilon x) + \frac{Q_0}{\epsilon} e^{-\epsilon x} \sin \epsilon x$$

$$+ M_L e^{-\epsilon(L-x)} [\sin \epsilon(L-x) + \cos \epsilon(L-x)]$$

$$- \frac{Q_L}{\epsilon} e^{-\epsilon(L-x)} \sin \epsilon(L-x) - \frac{1}{4\epsilon^4} \frac{d^2 p(x)}{dx^2} \qquad (5.44)$$

$$Q_x = -D_{11} \frac{d^3 w}{dx^3} = -2M_0\epsilon \, e^{-\epsilon x} \sin \epsilon x + Q_0 \, e^{-\epsilon x}(\cos \epsilon x - \sin \epsilon x)$$

$$+ 2M_L\epsilon \, e^{-\epsilon(L-x)} \sin \epsilon(L-x)$$

160

$$-Q_L\,e^{-\epsilon(L-x)}[-\cos\,\epsilon(L-x)+\sin\,\epsilon(L-x)]-\frac{1}{4\epsilon^4}\frac{d^3p(x)}{dx^3}$$

$$\text{(5.45)}$$

$N_x = \text{constant.}$ \hfill (5.46)

In the case of axially symmetric loading $\epsilon_\theta^0 = \kappa_\theta = 0$, the stresses in each lamina are given by

$$\begin{bmatrix} \sigma_x \\ \sigma_\theta \end{bmatrix}_k = \begin{bmatrix} Q_{11} & Q_{12} \\ Q_{12} & Q_{22} \end{bmatrix}_k \begin{bmatrix} \epsilon_x^0 \\ 0 \end{bmatrix} + \zeta \begin{bmatrix} Q_{11} & Q_{12} \\ Q_{12} & Q_{22} \end{bmatrix} \begin{bmatrix} \kappa_x \\ 0 \end{bmatrix} \qquad (5.47)$$

where $\epsilon_x^0 = \partial u_0/\partial x$, and $\kappa_x = -\partial^2 w/\partial x^2$. If the shell is of one orthotropic layer, u_0 is given by (5.39).

Waltz and Vinson [3] have shown that in a laminated cylindrical shell wherein $d^2p(x)/dx^2 = 0$, all interlaminar shear stresses go to zero outside of the bending boundary layer, that is, $4\sqrt{Rh(D_x/D_\theta)}^{1/2} \leqslant x \leqslant L - 4\sqrt{Rh(D_x/D_\theta)}^{1/2}$.

5.5. Response of a Long Axi-Symmetric Laminated Composite Shell to an Edge Displacement

Consider a long $(L > 4\sqrt{(D_x/D_\theta)}^{1/2}Rh)$ shell, composed of a laminated construction, $[0, \pm45°/90°]_S$, for two composite materials, subjected to an edge displacement $w(0) = w_0$, and $du(0)/dx = 0$ at one edge. Each of the eight plys are $h_k = 0.01''$, and the two materials are T300/5208 graphite epoxy and E-glass/epoxy with the following properties:

	T300/5208	E-glass/Epoxy
E_1	22.2×10^6 psi	8.8×10^6 psi
E_2	1.58×10^6 psi	3.6×10^6 psi
ν_{12}	0.30	0.23
ν_{21}	0.021	0.0941
G_{12}	0.81×10^6 psi	1.74×10^6 psi
v_f	0.70	0.72

Using the methods of Chapter 2, the components of the stiffness matrices are:

	$T300/5208$	$\cdot$	E-glass/Epoxy
A_{11}(lb./in.)	0.76×10^6		0.4667×10^6
A_{12}(lb./in.)	0.2355×10^6		0.1079×10^6
A_{22}(lb./in.)	0.76×10^6		0.4667×10^6
$A_{16}=A_{26}$	0		0
A_{66}(lb./in.)	0.2622×10^6		0.1794×10^6
B_{ij}	0	$\cdot$	0
D_{11}(lb.-in.)	6.74×10^2		3.1673×10^2

THE TERMS D_{12}, D_{22} ARE OMITTED FOR BREVITY AND $D_{16}=D_{26}=0$.

Stresses in each lamina are calculated at two axial locations to show the stress distributions that occur in the bending boundary layer.

Stresses are calculated at an axial location wherein $w=w_0/2$. Secondly, stresses are calculated at an axial location of $x=l/2$, where l is the axial distance where $w=0$, nearest to the end of the shell. Results are shown in Figure 5.3. It is seen that the σ_x stresses are tensile on the outer layers in general and compressive on the inner layers in general. It is also seen that the circumferential stresses, σ_y, are maximum in the $\theta=90°$ layers – which is logical because the circumferential stiffness is greatest in these layers and hence circumferential load is carried in the $90°$ layers analogous to the stiff spring in a collection of springs in parallel. Lastly, even though the deflections and stress are axially symmetric, in plane shear stresses, τ_{xy}, do exist because of the stacking sequence of the laminate.

5.6. Buckling of Circular Cylindrical Shells of Composite Materials Subjected to Various Loads.

5.6.1. Applied Loads

Consider a circular cylindrical shell shown in Figure 5.4 below of mean radius R, wall thickness h, and length L, subjected to a compressive load P, a beam-type bending moment M, a torque T, and an external pressure p. For ease of presentation the compression stress resultants resulting from some of these loads will be denoted as positive quantities, as opposed to the usual conventions

$$N_{x_{comp}} = \frac{P}{2\pi R} \tag{5.48}$$

$$\left(N_{x_{bend.}}\right)_{max} = \frac{M}{\pi R^2} \tag{5.49}$$

where N_x is the axial load per unit circumference.

In each of the above these are the applied loads. If these applied loads, the external pressure p, or the applied torque T, equal or exceed a critical

162

value, buckling will result, which for most practical purposes is synonymous to collapse and failure. The following is based on Reference 18.

5.6.2. Buckling

5.6.2.1. Buckling Due to Axial Compression
5.6.2.1.1. Assumptions
a. Special anisotropy [that is $(\ )_{16} = (\ )_{26} = 0$]

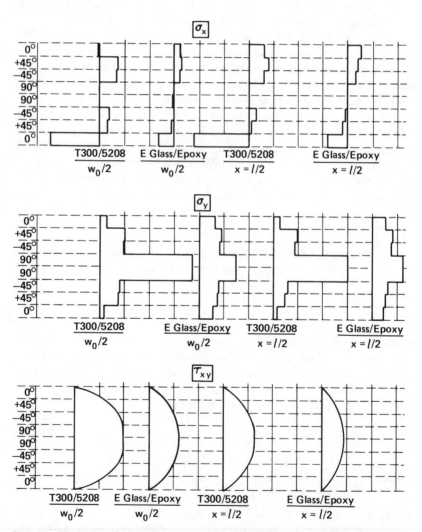

Figure 5.3. Stresses at selected points and/or displacements within the bending boundary layer.

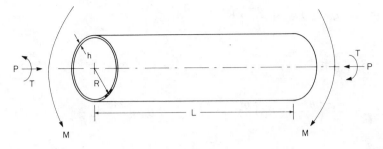

Figure 5.4.

b. Prebuckled deformations are not taken into account
c. Ends of the cylindrical shell are supported by rings rigid in their planes, but no resistance to rotation or bending out of their plane.

$sih \dfrac{m\,\mu x}{L} \cos \dfrac{n\,y}{R}$

$N_{c\ell} = N(m,n)$

  5.6.2.1.2. *General case, no mid plane symmetry, and n > 4.*

$$N_{x_{cr}} = \left(\frac{L}{m\pi}\right)^2 \frac{\begin{vmatrix} C_{11} & C_{12} & C_{13} \\ C_{21} & C_{22} & C_{23} \\ C_{31} & C_{32} & C_{33} \end{vmatrix}}{\begin{vmatrix} C_{11} & C_{12} \\ C_{21} & C_{22} \end{vmatrix}} \tag{5.50}$$

where
$N_{x_{cr}}$ critical compressive load per unit circumference
L cylinder length
R cylinder radius
m number of buckle *half waves* in the axial direction
n number of buckle *waves* in the circumferential direction

$$C_{11} = A_{11}\left(\frac{m\pi}{L}\right)^2 + A_{66}\left(\frac{n}{R}\right)^2 \tag{5.51}$$

$$C_{22} = A_{22}\left(\frac{n}{R}\right)^2 + A_{66}\left(\frac{m\pi}{L}\right)^2 \tag{5.52}$$

$$C_{33} = D_{11}\left(\frac{m\pi}{L}\right)^4 + D_{22}\left(\frac{m\pi}{L}\right)^2\left(\frac{n}{R}\right)^2 + D_{22}\left(\frac{n}{R}\right)^4$$

$$+ \frac{A_{22}}{R^2} + \frac{2B_{22}}{R}\left(\frac{n}{R}\right)^2 + \frac{2B_{12}}{R}\left(\frac{m\pi}{L}\right)^2 \tag{5.53}$$

$$C_{12} = C_{21} = (A_{12} + A_{66})\left(\frac{m\pi}{L}\right)\left(\frac{n}{R}\right) \tag{5.54}$$

$$C_{23} = C_{32} = \left(B_{12} + 2B_{66} \right)\left(\frac{m\pi}{L} \right)^2 \left(\frac{n}{R} \right) + \frac{A_{22}}{R}\left(\frac{n}{R} \right)$$

$$+ B_{22}\left(\frac{n}{R} \right)^3 \tag{5.55}$$

$$C_{13} = C_{31} = \frac{A_{12}}{R}\left(\frac{n\pi}{L} \right) + B_{11}\left(\frac{m\pi}{L} \right)^3 + \left(B_{12} + 2B_{66} \right)\left(\frac{m\pi}{L} \right)\left(\frac{n}{R} \right)^2 \tag{5.56}$$

$$A_{ij} = \sum_{k=1}^{N} \left[\overline{Q}_{ij} \right]_k \left(h_k - h_{k-1} \right) \tag{5.57}$$

$$B_{ij} = \tfrac{1}{2} \sum_{k=1}^{N} \left[\overline{Q}_{ij} \right]_k \left(h_k^2 - h_{k-1}^2 \right) \tag{5.58}$$

$$D_{ij} = \tfrac{1}{3} \sum_{k=1}^{N} \left[\overline{Q}_{ij} \right]_k \left(h_k^3 - h_{k-1}^3 \right) \tag{5.59}$$

It is seen that the A_{ij} comprise the extensional stiffness matrix of the composite shell, B_{ij} comprise the bending – extension coupling matrix, and the D_{ij} comprise the flexural or bending stiffness matrix of the shell material. All the notation is discussed in Chapter 2.

To determine the critical load $N_{x_{cr}}$ for a cylindrical shell with *given* dimensions and a *given* material system, one determines those integer values of m and n which make $N_{x_{cr}}$ a minimum. If choices can be made regarding the ply orientation and number of plys, then an *optimization* can be performed to determine the construction that provides the highest buckling load per unit weight.

After the buckling load has been determined, a check must be made to see that the final construction is not overstressed at a load below the critical buckling load, because if that is the case the cylinder is limited to a load that will result in overstressing. For a laminated composite construction, the most general constitutive equation for the cylinder, is given by:

$$\begin{bmatrix} N_x \\ N_\theta \\ N_{x\theta} \\ \hline M_x \\ M_\theta \\ M_{x\theta} \end{bmatrix} = \left[\begin{array}{ccc|ccc} A_{11} & A_{12} & 2A_{16} & B_{11} & B_{12} & 2B_{16} \\ A_{12} & A_{22} & 2A_{26} & B_{12} & B_{22} & 2B_{26} \\ A_{16} & A_{26} & 2A_{66} & B_{16} & B_{26} & 2B_{66} \\ \hline B_{11} & B_{12} & 2B_{16} & D_{11} & D_{12} & 2D_{16} \\ B_{12} & B_{22} & 2B_{26} & D_{12} & D_{22} & 2D_{26} \\ B_{16} & B_{26} & 2B_{66} & D_{16} & D_{26} & 2D_{66} \end{array} \right] \begin{bmatrix} \epsilon_x^0 \\ \epsilon_\theta^0 \\ \epsilon_{x\theta}^0 \\ \hline \kappa_x \\ \kappa_\theta \\ \tfrac{1}{2}\kappa_{x\theta} \end{bmatrix} \tag{5.60}$$

In the case of buckling due to axial compression, $N_x = N_{x_{cr}}$, $N_\theta = N_{x\theta} = M_x = M_\theta = M_{x\theta} = 0$ we obtain the $[\epsilon^0]$ and $[\kappa]$ matrices from the above. Then, using those matrices each stress component σ_x, σ_θ, $\sigma_{x\theta}$, in each lamina and ply can be calculated by

$$
\begin{bmatrix} \sigma_x \\ \sigma_\theta \\ \sigma_{x\theta} \end{bmatrix}_k = [\overline{Q}]_k \begin{bmatrix} \epsilon_x^0 \\ \epsilon_\theta^0 \\ \epsilon_{x\theta}^0 \end{bmatrix} + z[\overline{Q}]_k \begin{bmatrix} \kappa_x \\ \kappa_\theta \\ \tfrac{1}{2}\kappa_{x\theta} \end{bmatrix}
\tag{5.61}
$$

These stresses can then be compared to the allowable or failure stress in each ply as discussed in Chapter 7.

5.6.2.1.3. *Special Case, Mid Plane Symmetry [that is $B_{ij} = 0$]*

For this case the equation to use is

$$
\frac{N_{x_{cr}} L^2}{\pi^2 D_{11}} = m^2 \left(1 + 2\frac{D_{12}}{D_{11}}\beta^2 + \frac{D_{22}}{D_{11}}\beta^4 \right)
$$

$$
+ \frac{\gamma^2 L^4}{\pi^4 m^2 D_{11} R^2} \cdot \frac{A_{11}A_{22} - A_{12}^2}{A_{11} + \left(\dfrac{A_{11}A_{22} - A_{12}^2}{A_{66}} - 2A_{12} \right)\beta^2 + A_{22}\beta^4}
\tag{5.62}
$$

where

$$
\beta = \frac{nL}{\pi R m}
\tag{5.63}
$$

$$
\gamma = 1.0 - 0.901(1 - e^{-\phi})
\tag{5.64}
$$

$$
\phi = \frac{1}{29.8} \left[\frac{R}{\sqrt{\dfrac{D_{11}D_{22}}{A_{11}A_{22}}}} \right]^{1/2}
\tag{5.65}
$$

Here γ is an empirical (knock down) factor that insures that the calculated buckling load will be conservative with respect to all experimental data that are available.

To determine the critical buckling load $N_{x_{cr}}$, vary the *integers m and n* to determine the minimum value of $N_{x_{cr}}$, which will be the actual buckling load. Again one may perform an *optimization* through a calculation to insure that overstressing of some ply or plys has not occurred at a load below the critical buckling load, using (5.60) and (5.61) above.

5.5.2.2. Buckling Due to Bending of the Cylindrical Shell (Symmetrical Case $B_{ij} = 0$)

The equations to use are Equations (5.62), (5.63) and (5.65) wherein the empirical factor γ is given by:

$$\gamma = 1 - 0.731 \left(1 - e^{-\phi}\right) \tag{5.66}$$

The procedures to find the critical buckling load $N_{x_{cr}}$ for a given geometry and construction, the procedures to *optimize* the construction, and the procedures to insure there is no *overstressing* are identical to the procedures of Section 5.5.2.1.

5.6.2.3. Buckling Due to External Lateral Pressure and Hydrostatic Pressure.

For a lateral external pressure, the critical value of pressure that will cause buckling is determined by:

$$p_{cr} = \frac{R}{n^2} \frac{\begin{vmatrix} C_{11} & C_{12} & C_{13} \\ C_{21} & C_{22} & C_{23} \\ C_{31} & C_{32} & C_{33} \end{vmatrix}}{\begin{vmatrix} C_{11} & C_{12} \\ C_{21} & C_{22} \end{vmatrix}} \tag{5.67}$$

In this case $m = 1$, and one varies the integer n ($n \geqslant 2$) to find the minimum value of p_{cr} for a given construction because that is the physical buckling load.

For long cylinders subjected to a lateral pressure, the critical buckling pressure is given by

$$p_{cr} = \frac{3\left(D_{22} - \dfrac{B_{22}^2}{A_{22}}\right)}{R^3} \tag{5.68}$$

In the case where $B_{11} = B_{22} = B_{12} = B_{66} = 0$, p_{cr} is found by the following,

$$p_{cr} = \frac{5.513}{LR^{3/2}} \left[\frac{D_{22}^3\left(A_{11}A_{22} - A_{12}^2\right)}{A_{22}} \right]^{1/4} \tag{5.69}$$

which is valid only when

$$\left(\frac{D_{22}}{D_{11}}\right)^{3/2} \left(\frac{A_{11}A_{22} - A_{12}^2}{12A_{22}D_{11}}\right)^{1/2} \left(\frac{L^2}{R}\right) > 500 \tag{5.70}$$

If the external pressure is hydrostatic, use (5.67) but replace n^2 by

$$n^2 + \frac{1}{2}\left(\frac{m\pi R}{L}\right)^2.$$ (5.71)

One varies the integers m and n to obtain the combination that gives the lowest (physical) buckling pressure that will occur.
NOTE: In all cases above it is recommended that the calculated critical pressure, p_{cr}, be multiplied by 0.75 for use in *design*.
Again if this number of plys can be varied as well as their orientation then an *optimization* can be made to obtain the highest buckling pressure per unit weight.
To insure that overstressing does not occur at a pressure lower than the critical pressure:
For the external lateral pressure only:

$$N_\theta = pR, \quad N_x = N_{x\theta} = M_x = M_\theta = M_{x\theta} = 0$$ (5.72)

For a hydrostatic pressure:

$$N = pR, \ N_x = p\frac{R}{2}, \quad N_{x\theta} = M_x = M_\theta = M_{x\theta} = 0$$ (5.73)

Use these in Eq. (5.60) to obtain the $[\epsilon^0]$ and $[\kappa]$ matrix, substitute those into (5.61) to get each stress in each ply, and compare that stress to an allowable or failure stress, discussed in Chapter 7, to insure that no overstressing occurs.

5.6.2.4. Buckling Due to a Torsional Load.
When $[B_{ij} \approx 0]$, the critical torque T_{cr} that can be applied is:

$$T_{cr} = 21.75(D_{22})^{5/8}\left(\frac{A_{11}A_{22} - A_{12}^2}{A_{22}}\right)^{3/8}\frac{R^{5/4}}{L^{1/2}}.$$ (5.74)

Another restriction on this equation is that

$$\left(\frac{D_{22}}{D_{11}}\right)^{5/6}\left(\frac{A_{11}A_{22} - A_{12}^2}{12A_{22}D_{11}}\right)^{1/2}\frac{L^2}{R} \geqslant 500$$ (5.75)

It is recommended that the T_{cr} determined by (5.74) above be multiplied by 0.67 for use in *design*.
T_{cr} is obtained from (5.74) directly. An optimized structure can be obtained by varying the θ's in each ply as well as the number of plys to obtain a construction that is minimum weight for a given T_{cr}. To insure

that there is no overstressing in any ply at T_{cr} in equation (5.60) insert

$$N_{x\theta} = \frac{T_{cr}}{2\pi R}, \quad N_x = N_\theta = M_x = M_\theta = M_{x\theta} = 0$$

and determine the $[\epsilon^0]$ and the $[\kappa]$ matrices, which in turn, upon substitution into (5.61) gives the stresses in each ply, to be compared to allowable or fracture values, discussed in Chapter 7.

5.6.2.5. Buckling Due to Combined Axial Compression and Bending

The interaction equations for this combined loading is

$$R_c + R_B = 1$$

$$\text{where } R_c = \frac{N_{x\,comp}}{N_{x_{cr}}}, \quad R_B = \frac{N_{x\,bend}}{N_{x_{cr}}}. \tag{5.76}$$

For the applied loads in the numerators of (5.76), use (5.48) for axial compression and (5.49) for bending. The critical values are obtained from Sections 5.5.2.1 and 5.5.2.2 respectively.

If $R_c + R_B < 1$, the constructions will not buckle.

For a given construction, the integer m and n must both be varied simultaneously for both loading such that the left hand side of (5.76) is a maximum for a given set of applied loads. Also a combined optimization can be made to find a construction that will be minimum weight for a given set of loads. To check overstressing Equations (5.60) and (5.61) are again employed where the only non zero load is

$$N_x = \frac{P_{applied}}{2\pi R} + \frac{M_{applied}}{\pi R^2}.$$

5.6.2.6. Buckling Due to Combined Axial Compression and External Pressure.

The interactive equation to employ, although it is not comletely established, is:

$$R_c + R_p = 1 \tag{5.77}$$

where $R_p = p_{applied}/p_{cr}$, and R_c is given previously in (5.76), and p_{cr} is determined from the equations of Section 5.5.2.3 for either an external lateral pressure or a hydrostatic pressure.

Utilize the statements of previous Sections to analyze the adequacy of the shell for given loads, *optimize* the shell for given loads, and for overstressing calculations utilize Equations (5.60) and (5.61) where for a combination of axial compression and external lateral pressure

$$N_x = \frac{P_{\text{applied}}}{2\pi R}, \quad N_\theta = p_{\text{app.}} R, \quad \text{all other loads} = 0$$

for axial compression and hydrostatic pressure

$$N_x = \frac{P_{\text{applied}}}{2\pi R} + \frac{p_{\text{app.}} R}{2}, \quad N_\theta = p_{\text{app.}} R, \text{ all other loads} = 0.$$

5.6.2.7. Buckling Due to Axial Compression and a Torsional Load
The interaction equations to utilize, although its still preliminary is

$$R_c + R_T = 1 \tag{5.78}$$

when $R_T = \dfrac{T_{\text{applied}}}{T_{\text{cr}}}$

where T_{cr} is taken from (5.74). The procedure to follow is equivalent to that of Section 5.5.2.6, but in checking for overstressing. The non zero loads to employ with Equations (5.60) and (5.61) are:

$$N_x = \frac{P_{\text{app.}}}{2\pi R}, \quad N_{x\theta} = \frac{T_{\text{app.}}}{2\pi R}.$$

5.7. Vibration of Composite Shells.

The free vibration of laminated orthotropic cylindrical shells using classical theory has been studied by White [19], Dong [20], Bert [21], and Tasi [22,23] and Tasi and Roy [24]. Mirsky [25,26] included transverse shear deformation and transverse normal strain of orthotropic homogeneous cylindrical shells. Dong and Tso [27] analyzed the vibration of layered orthotropic cylinders including transverse shear deformation. Sun and Whitney [28] studied the axially symmetric vibration of laminated composite cylindrical shell including transverse shear deformation, transverse normal stress and strain, rotatory inertia and higher order stiffness and inertia terms. The work of Bert [15] is mentioned here again.

The analytical and experimental aspects of dynamic response of composite shells have been studied by Ross, Sierakowski and Sun [29], including various experimental techniques and models.

5.8. Problems and Exercises

5.1. Consider a cylindrical shell composed of four laminae of T300/5208 unidirectional graphite-epoxy whose properties include:

$E_1 = 21 \times 10^6$ psi $\nu_{12} = 0.21, \quad \nu_{21} = 0.017$

$E_2 = 1.76 \times 10^6$ psi $\rho = 0.06$ lbs/in^3, $\quad g = 386$ in/sec^2

$G_{12} = 0.65 \times 10^6$ psi $\sigma_{allowable} = 100,000$ psi

If the plies are 0.0055" thick, all oriented with $\theta = 0°$, (that is, unidirectional), and if the radius of the shell is $R = 12"$
(a) What is D_{11}?
(b) What is D_{22}?
(c) What is the length of the bending boundary layer at each end?

5.2. Consider a circular cylindrical shell of length $L = 50"$, radius $R = 10"$, and a wall thickness $h = 0.020"$, composed of a unidirectional composite of properties $D_x = D_{11} = 28.053$ lb-in, and $D_\theta = D_{22} = 0.5874$ lb-in. What is the length of the bending boundary layer?

5.3. For a circular cylindrical shell, composed of the unidirectional graphite epoxy given in Problem 5.1 for a shell of radius 15", wall thickness of 0.15" and length 30", what is the length of the bending boundary layer if:
(a) The fibers are in the axial direction?
(b) The fibers are in the circumferential direction?

5.4. Consider the cylindrical shell of Problem 5.3 with the material properties given in Problem 5.1, where the fibers are in the axial direction. If the shell is clamped at both ends and subjected to $p = 100$ psi and $N_x = 0$,
(a) What are σ_x and σ_θ at $x = 0$?
(b) What are σ_x and σ_θ at $x = 15"$ (i.e. $x = L/2$)?

5.5. Consider a circular cylindrical shell of length $L = 50"$, radius $R = 10"$, and wall thickness $h = 0.020"$, composed of the same material as in Problem 5.1 wherein $D_x = D_{11}$ and $D_\theta = D_{22}$. What is the length of the bending boundary layer for this shell?

5.6. Consider a shell composed of a Quasi-isotropic composite material wherein $E_x = E_\theta = 17 \times 10^6$ psi and $\nu = 0.3$. The mean shell radius is 10 inches and the total thickness is $h = 0.3$ inches. The length is 40 inches, and the interlaminar shear strength sufficient.
(a) What is the length of the bending boundary layer?
(b) What is the membrane stress σ_x in the shell if it were subjected to a tensile axial load of 10^4 lbs?

5.9. References

1. Vinson, J.R. and T.W. Chou. *Composite Materials and Their Use in Structures,* Applied Science Publishers, London. (1975).
2. Wilkins, D.J. and T.S. Love. "Combined Compression – Torsion Buckling Tests of Laminated Composite Cylindrical Shells." *AIAA Journal of Aircraft, v. 12, No. 11.* (Nov. 1975): 885–889.
3. Waltz, T. and J.R. Vinson, "Interlaminar Stresses in Laminated Cylindrical Shells of Composite Materials." *AIAA Journal, Vol. 14, No. 9.* (September 1976): 1213–1218.
4. El Naschie, M.S. "Initial and Post Buckling of Axially Compressed Orthotropic Cylindrical Shells." *AIAA Journal v. 14, No. 10* (October 1976): 1502–1504.
5. Ecord, G.M. "Composite Pressure Vessels for the Space Shuttle Orbiter." *Composites in Pressure Vessels and Pipings, ASME – PVP-BB-021* (1977): 129–140.
6. Johnson, R., R.J. Reck, and R.C. Davis. "Design and Fabrication of a Large Graphite-Epoxy Cylindrical Shell." *Proceedings AIAA/ASME 19th Structures, Structural Dynamic and Materials Conference* (1978): 300–310.
7. Fujczak, R.R. "Torsional Fatigue Behavior of Graphite-Epoxy Cylinders." *Proceedings of the 2nd International Conference on Composite Materials (ICCM2)* (1978): 635–648.
8. Booton, M. and R.C. Tennyson, "Buckling of Imperfect Anisotropic Circular Cylinders Under Combined Loading." *Proceedings AIAA/ASME 19th Structures, Structural Dynamics, Materials Conference* (1978): 351–358.
9. Raju, B.B., R. Chandra, and M.S. Rao. "Transient Temperatures in Laminated Composite Conical Shells Due to Aerodynamic Heating." *AIAA Journal, Vol. 6, No. 6* (June 1978): 547–548.
10. Varadan, T.K., "Snap-Buckling of Orthotropic Shallow Spherical Shells." *Journal of Applied Mechanics, Vol. 45, No. 2* (June 1978): 445–447.
11. Rhodes, J. and I.H. Marshall." Unsymmetrical Buckling of Laterally Loaded Reinforced Plastic Shells." *Proceedings of the 2nd International Conference of Composite Materials (ICCM2),* (1978): 303–315.
12. Montague, P. "Experimental Behavior of Double-Skinned, Composite, Circular Cylindrical Shells Under External Pressure." *Journal of Mechanical Engineering Science, vol. 20, No. 1* (1978): 21–34.
13. Humphrey, W.D. "Degradation Data of Kevlar Pressure Vessels", *NBS Publication 563* (1979): 177–185.
14. Bert, C.W. and V.S. Reddy. "Cylindrical Shells of Bimodulus Composite Material." *Oklahoma University Report OU-AMNE-80-3* (February 1980).
15. Bert, C.W. and M. Kumar. "Vibration of Cylindrical Shells of Bimodulus Composite Materials." *Oklahoma University Report OU-AMNE-80-20.* (October 1980).
16. Yuceoglu, U. and D.P. Updike. "Stress Concentration in Bonded Multi-Layer Cylindrical Shells." *Journal of Engineering Science* (1981).
17. Vinson, J.R. *"Structural Mechanics: The Behavior of Plates and Shells."* John Wiley and Sons (1974).
18. Anon. "Buckling of Thin Walled Circular Cylinders," NASA SP-8007. Revised, August, 1968.
19. White, J.C., "The Flexural Vibrations of Thin Laminated Cylinders." Journal of Engineering Industry (1961): 397–402.
20. Dong, S.B. "Free Vibration of Laminated Orthotropic Cylindrical Shells." *Journal of the Acoustical Society of America, vol. 44* (1968): 1628–1635.
21. Bert, C.W., V. Baker and D. Egle, "Free Vibrtions of Multilayer Anisotropic Cylindrical Shells, *Journal of Composite Materials, vol. 3* (1969): 480–499.
22. Tasi, J. "Reflection of Extensional Waves at the End of a Thin Cylindrical Shell," *Journal of the Acoustical Society of America, vol. 44,* (1968): 291–292.

172

23. Tasi, J. "Effect of Heterogeneity on the Axisymmetric Vibration of Cylindrical Shells," *Journal of Sound and Vibration, vol. 14* (1971) 325–328.

24. Tasi, J. and Roy, B.N. "Axisymmetric Vibration of Finete Heterogeneous Cylindrical Shells," Journal of Sound and Vibration, vol. 17, (1971): 83–94.

25. Mirsky, I. "Vibration of Orthotropic, Thick, Cylindrical Shells," *Journal of the Acoustical Society of America, vol. 36* (1964): 41–51.

26. Mirsky, I. Axisymmetric Vibrations of Orthotropic Cylinders, *Journal of the Acoustic Society of America, vol. 36* (1964): 2106–2112.

27. Dong, S.B. and Tso, "On a Laminated Orthotropic Shell Theory Including Transverse Shear Deformation", *Journal of Applied Mechanics, vol. 39* (1972): 1091–1097.

28. Sun, C.T. and Whitney, J.M. "Axisymmetric Vibrations of Laminated Composite Cylindrical Shells," *Journal of the Acoustical Society of America, vol. 55* (June 1974): 1238–1246.

29. Ross, C.A., R.L. Sierakowski and C.T. Sun. "Dynamic Response of Composite Materials," Society of Experimental Mechanics Publication S-014, 1980.

6. Energy Methods in Composite Material Structures

6.1. Introduction

Many composite material structures not only involve anisotropy, multilayer considerations and transverse shear deformation, but also have hygrothermal effects, which can be very important. True, for preliminary design one needs the simplified, easier to use analyses that have been presented earlier, but for *the* final design, transverse shear deformation and hygrothermal effects must be included. These thermal and moisture effects have been described in Chapter 2. Analytically they cause considerable difficulty, because with their inclusion few boundary conditions are homogeneous, hence separation of variables, used throughout the plate and shell solutions to this point, cannot be utilized straightforwardly. Only through the laborious process of transformation of variables can the procedure used herein be used [1]. Therefore, energy principles are much more convenient for use in design and analyses of plate and shell structures when hygrothermal effects are present.

There are three energy principles used in structural mechanics. The Theorem of Minimum Potential Energy, The Theorem of Minimum Complementary Energy and Reissner's Variational Theorem. The latter has been used extensively in an earlier text [1], and could be re-used here. Minimum Complementary Energy is seldom used. The Theorem of Minimum Potential Energy will be described in this chapter and utilized effectively in what follows because it effectively treats *all* problems involving all the effects needed for the detailed, accurate analysis and design of composite material structures.

As stated just above, an alternative to deriving the governing differential equations for an elastic body such as a beam, plate or shell, and solving them to satisfy prescribed boundary conditions, is to develop expressions involving the strain energy of the body and the work done by forces acting on the structures. The Theorem of Minimum Potential Energy method will be introduced here, because it is that technique that was used to solve two major in-depth studies that will be treated in subsequent sections.

6.2. Theorem of Minimum Potential Energy.

For any generalized elastic body, the potential energy of that body can be written as follows:

$$V = \int_R W \, dR - \int_{S_T} T_i U_i \, dS - \int_R F_i U_i \, dR \tag{6.1}$$

where

W strain energy density function, defined in (6.4) below
R volume of the elastic body
T_i i^{th} component of the surface traction
U_i i^{th} component of the deformation
F_i i^{th} component of a body force
S_T portion of the body surface over which tractions are prescribed

It is seen that the first term on the right hand side of (6.1) is the strain energy of the elastic body. The second and third terms are the work done by the surface tractions and the body forces, respectively.

The Theorem of Minimum Potential Energy can be stated as:

"Of all of the displacements satisfying compatibility and the prescribed boundary conditions, those which satisfy the equilibrium equations make the potential energy a minimum."

Mathematically, the operation is simply

$$\delta V = 0. \tag{6.2}$$

The lower case delta is a mathematical operation known as a variation. Operationally, it is analogous to partial differentiation. To employ variational operations in structural mechanics only the following three operations are needed:

$$\frac{d(\delta y)}{dx} = \delta\left(\frac{dy}{dx}\right), \quad \delta(y^2) = 2y\delta y, \quad \int \delta y \, dx = \delta \int y \, dx \tag{6.3}$$

In (6.1) the strain energy density function, W, is defined as follows in a Cartesian coordinate frame:

$$W = \tfrac{1}{2}\sigma_{ij}\epsilon_{ij} = \tfrac{1}{2}\sigma_x\epsilon_x + \tfrac{1}{2}\sigma_y\epsilon_y + \tfrac{1}{2}\sigma_z\epsilon_z$$

$$+ \sigma_{xy}\epsilon_{xy} + \sigma_{xz}\epsilon_{xz} + \sigma_{yz}\epsilon_{yz} \tag{6.4}$$

To utilize the Theorem of Minimum Potential Energy, the stress strain relations for the elastic body are employed to change the stresses in (6.4) to strains, and the strain displacement relations are employed to change all strains to displacements.

6.3. Beam Analysis

As an example of the use of minimum potential energy consider the following beam in bending, shown in Figure 6.1.

From Figure 6.1 it is seen the beam is of length L, in the x-direction, width b and height h. It is subjected to a lateral distributed load, $q(x)$ in the z-direction, in units of lbs./in. of beam length. The modulus of elasticity of the isotropic beam material is E, and the stress strain relations are

$$\sigma_x = E\epsilon_x \tag{6.5}$$

and the strain displacement relations are

$$\epsilon_x = \frac{du_x}{dx} = -\frac{d^2w}{dx^2}z \tag{6.6}$$

since the bending of beams, $u_x = -z(\partial w/\partial x)$ only.

Looking at $(6.4)-(6.6)$ and remembering that in elementary beam theory $\sigma_y = \sigma_z = \sigma_{xy} = \epsilon_{xz} = \epsilon_{yz} = \sigma_{yz} = 0$, then

$$W = \tfrac{1}{2}\sigma_x\epsilon_x = \tfrac{1}{2}E\epsilon_x^2 = \tfrac{1}{2}E\left(\frac{d^2w}{dx^2}\right)^2 z^2 \tag{6.7}$$

Therefore, the strain energy, U, is:

$$U = \int_0^L \int_{-b/2}^{+b/2} \int_{-h/2}^{+h/2} \tfrac{1}{2}E\left(\frac{d^2w}{dx^2}\right)^2 z^2 \, dz \, dy \, dx$$

$$= \frac{EI}{2}\int_0^L \left(\frac{d^2w}{dx^2}\right)^2 dx \tag{6.8}$$

where $I = bh^3/12$

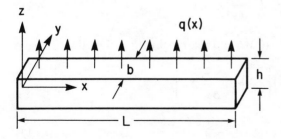

Fig. 6.1.

Similarly, from the surface traction force term in (6.1) it is seen that

$$\int_{S_T} T_i U_i ds = \int_0^L q(x) w(x) dx$$

Equation (6.1) then becomes

$$V = \frac{EI}{2} \int_0^L \left(\frac{d^2 w}{dx^2} \right)^2 dx - \int_0^L q(x) w(x) dx \tag{6.9}$$

Following (6.2) and remembering (6.3) then

$$\delta V = 0 = EI \int_0^L \delta \left(\frac{d^2 w}{dx^2} \right)^2 dx - \int_0^L q(x) \delta w(x) dx \tag{6.10}$$

The δ's can be included under the integral, because the order of variation and integration can be interchanged. Also, there is no variation of E, I or $q(x)$ because they are all specified quantities.

From the first term on the right hand side of (6.10) and integrating by parts:

$$\frac{EI}{2} \int_0^L \delta \left(\frac{d^2 w}{dx^2} \right)^2 dx = EI \int_0^L \left(\frac{d^2 w}{dx^2} \right) \delta \left(\frac{d^2 w}{dx^2} \right) dx$$

$$= EI \int_0^L \frac{d^2 w}{dx^2} \frac{d^2(\delta w)}{dx^2} dx$$

$$= \left[EI \frac{d^2 w}{dx^2} \delta \left(\frac{dw}{dx} \right) \right]_0^L - EI \int_0^L \frac{d^3 w}{dx^3} \frac{d(\delta w)}{dx}$$

$$= \left[EI \frac{d^2 w}{dx^2} \delta \left(\frac{dw}{dx} \right) \right]_0^L - \left[EI \frac{d^3 w}{dx^3} \delta w \right]_0^L$$

$$- EI \int_0^L \frac{d^4 w}{dx^4} \delta w dx \tag{6.11}$$

Substituting (6.11) into (6.10) and rearranging, it is seen that:

$$\delta V = 0 = \left[EI \frac{d^2 w}{dx^2} \delta \left(\frac{dw}{dx} \right) \right]_0^L - \left[EI \frac{d^3 w}{dx^3} \delta w \right]_0^L$$

$$+ \int_0^L \left[EI \frac{d^4 w}{dx^4} - q(x) \right] \delta w dx = 0 \tag{6.12}$$

For this to be true, the following equation must be satisfied for the integral above to be zero:

$$EI\frac{d^4 w}{dx^4} = q(x) \tag{6.13}$$

This is obviously the governing equation for the bending of a beam under a lateral load. It could have been derived in other ways. However, it lucidly illustrates the point that if a nonclassical shaped elastic structure were being analyzed, by using physical intuition, experience, or some other reason, one can formulate stress-strain relations, and strain displacement relations for the body, and through the Theorem can formulate equations analogous to (6.13). Incidentally, the resulting governing equations evolved from the Theorem of Minimum Potential Energy are called the Euler-Lagrange equations.

Note also for (6.12) to be true it is seen that each of the first two terms must be zero also. Hence, at $x = 0$ and $x = L$, (at each end) either $EI(d^2 w/dx^2) = -M_x = 0$ or (dw/dx) must be specified (that is, its variation must be zero), also either $EI(d^3 w/dx^3)(= -V_x) = 0$ or w must be specified.

It is seen that all of the classical boundary conditions including simple supports, clamped and free edges are all contained in the above "natural boundary conditions," which are a nice byproduct from using the variational approach to deriving governing equations for analyzing any elastic structure.

6.4. A Rectangular Composite Material Plate Subjected to Lateral and Hygrothermal Loads

6.4.1. Formulation of the Potential Energy Theorem

A detailed study by Sloan [3] illustrates clearly what is involved in analyzing composite panels to accurately account for the effects of anisotropy, transverse shear deformation, thermal and hygrothermal effects.

The stress strain equations to be considered are given by Equation (2.36), wherein the $\overline{Q}_{ij}$ are given on pages 47 and 48. The strain-displacement equations are given by (2.40), the form of the panel displacements by (2.41), from which (2.42) results. By neglecting ϵ_z and σ_z, the constitutive relations for the laminate reduce to (2.62). Sloan has shown that even for this problem, those quantities can be neglected.

Employing the Theorem of Minimum Potential Energy, Equation (6.1), for the plate under discussion it is seen that upon summing all the strain energy density functions for each lamina across the N laminae comprising the plate are given by

$$V = \tfrac{1}{2} \sum_{k=1}^{N} \iint_A \int_{h_{k-1}}^{h_k} \left\{ \sigma_x [\epsilon_x - \alpha_x \Delta T - \beta_x \Delta m] \right.$$

$$+ \sigma_y \left[\epsilon_y - \alpha_y \Delta T - \beta_y \Delta m \right] + \sigma_{xz} [2\epsilon_{xz}]$$

$$\left. + \sigma_{yz} [2\epsilon_{yz}] + \sigma_{xy} \left[2\epsilon_{xy} - \alpha_{xy} \Delta T - \beta_{xy} \Delta m \right] \right\} dz\, dA$$

$$- \iint_A p(x, y) w(x, y)\, dA. \tag{6.14}$$

Here A refers to the planform area of the plate whose dimensions are $0 \leqslant x \leqslant a$, $0 \leqslant y \leqslant b$ and $-h/2 \leqslant z \leqslant h/2$. It is noted that the strains used in the strain energy relations are the isothermal strains, hence one sees the differences between total strain and the thermal and hygrothermal strains in (6.14)

Now, substituting the constitutive equations (2.62) and the strain-displacement relations (2.40) into (6.14) there results:

$$V = \iint_A \left\{ \frac{A_{11}}{2} \left(\frac{du_0}{dx} \right)^2 + B_{11} \frac{\partial u_0}{\partial x} \frac{\partial \bar{\alpha}}{\partial x} + \frac{D_{11}}{2} \left(\frac{\partial \bar{\alpha}}{\partial x} \right)^2 + A_{12} \frac{\partial u_0}{\partial x} \frac{\partial v_0}{\partial y} \right.$$

$$+ B_{12} \left[\frac{\partial u_0}{\partial x} \frac{\partial \bar{\beta}}{\partial y} + \frac{\partial v_0}{\partial y} \frac{\partial \bar{\alpha}}{\partial x} \right] + D_{12} \frac{\partial \bar{\beta}}{\partial y} \frac{\partial \bar{\alpha}}{\partial x} + A_{16} \left[\frac{\partial u_0}{\partial x} \frac{\partial u_0}{\partial y} + \frac{\partial u_0}{\partial x} \frac{\partial v_0}{\partial x} \right]$$

$$+ D_{16} \left[\frac{\partial \bar{\alpha}}{\partial x} \frac{\partial \bar{\alpha}}{\partial y} + \frac{\partial \bar{\alpha}}{\partial x} \frac{\partial \bar{\beta}}{\partial x} \right] + B_{16} \left[\frac{\partial u_0}{\partial x} \frac{\partial \bar{\alpha}}{\partial y} + \frac{\partial u_0}{\partial x} \frac{\partial \bar{\beta}}{\partial x} + \frac{\partial u_0}{\partial y} \frac{\partial \bar{\alpha}}{\partial x} + \frac{\partial v_0}{\partial x} \frac{\partial \bar{\alpha}}{\partial x} \right]$$

$$+ \frac{A_{22}}{2} \left(\frac{\partial v_0}{\partial y} \right)^2 + B_{22} \frac{\partial v_0}{\partial y} \frac{\partial \bar{\beta}}{\partial y} + \frac{D_{22}}{2} \left(\frac{\partial \bar{\beta}}{\partial y} \right)^2 + A_{26} \left[\frac{\partial v_0}{\partial y} \frac{\partial u_0}{\partial y} + \frac{\partial v_0}{\partial y} \frac{\partial v_0}{\partial x} \right]$$

$$+ B_{26} \left[\frac{\partial v_0}{\partial x} \frac{\partial \bar{\alpha}}{\partial y} + \frac{\partial v_0}{\partial y} \frac{\partial \bar{\beta}}{\partial x} + \frac{\partial u_0}{\partial y} \frac{\partial \bar{\beta}}{\partial y} + \frac{\partial v_0}{\partial x} \frac{\partial \bar{\beta}}{\partial y} \right] + D_{26} \left[\frac{\partial \bar{\alpha}}{\partial y} \frac{\partial \bar{\beta}}{\partial y} + \frac{\partial \bar{\beta}}{\partial x} \frac{\partial \bar{\beta}}{\partial y} \right]$$

$$+ A_{45} \left[\bar{\alpha}\bar{\beta} + \bar{\alpha} \frac{\partial w}{\partial y} + \bar{\beta} \frac{\partial w}{\partial x} + \frac{\partial w}{\partial x} \frac{\partial w}{\partial y} \right] + A_{55} \left[\frac{\bar{\alpha}^2}{2} + \alpha \frac{\partial w}{\partial x} + \frac{1}{2} \left(\frac{\partial w}{\partial x} \right)^2 \right]$$

$$+ A_{44} \left[\frac{\bar{\beta}^2}{2} + \beta \frac{\partial w}{\partial y} + \frac{1}{2} \left(\frac{\partial w}{\partial y} \right)^2 \right] + A_{66} \left[\frac{1}{2} \left(\frac{\partial u_0}{\partial y} \right)^2 + \frac{\partial u_0}{\partial y} \frac{\partial v_0}{\partial x} + \frac{1}{2} \left(\frac{\partial v_0}{\partial x} \right)^2 \right]$$

$$+ B_{66} \left[\frac{\partial u_0}{\partial y} \frac{\partial \bar{\alpha}}{\partial y} + \frac{\partial u_0}{\partial y} \frac{\partial \bar{\beta}}{\partial x} + \frac{\partial v_0}{\partial x} \frac{\partial \bar{\alpha}}{\partial y} + \frac{\partial v_0}{\partial x} \frac{\partial \bar{\beta}}{\partial x} \right]$$

$$+ D_{66} \left[\frac{1}{2} \left(\frac{\partial \bar{\alpha}}{\partial y} \right)^2 + \frac{\partial \bar{\alpha}}{\partial y} \frac{\partial \bar{\beta}}{\partial x} + \frac{1}{2} \left(\frac{\partial \bar{\beta}}{\partial x} \right)^2 \right] - \frac{\partial u_0}{\partial x} \left(N_{1j}^T + N_{1j}^m \right)$$

$$- \frac{\partial v_0}{\partial y} \left(N_{2j}^T + N_{2j}^m \right) - \left(\frac{\partial u_0}{\partial y} + \frac{\partial v_0}{\partial x} \right) \left(N_{6j}^T + N_{6j}^m \right) - \frac{\partial \alpha}{\partial x} \left(M_{1j}^T + M_{1j}^m \right)$$

$$- \frac{\partial \bar{\beta}}{\partial y} \left(M_{2j}^T + M_{2j}^m \right) - \left(\frac{\partial \alpha}{\partial y} + \frac{\partial \beta}{\partial x} \right) \left(M_{6j}^T + M_{6j}^m \right)$$

$$+ T^* + M^* + \overline{MT^*} - p(x, y)w(x, y) \Big\} \mathrm{d}A \tag{6.15}$$

where all quantities except displacements are defined by Equations (2.50) through (2.53), (2.55) through (2.57) and

$$T^* = \frac{1}{2} \sum_{k=1}^{N} \int_{h_{k-1}}^{h_k} \left[\overline{Q}_{ij} \right]_k [\alpha_i]_k [\alpha_j]_k \Delta T^2(z, t) \mathrm{d}z$$

$$M^* = \frac{1}{2} \sum_{k=1}^{N} \int_{h_{k-1}}^{h_k} \left[\overline{Q}_{ij} \right]_k [\beta_i]_k [\beta_j]_k \Delta m^2(z, t) \mathrm{d}z$$

$$\overline{MT^*} = \sum_{k=1}^{N} \int_{h_{k-1}}^{h_k} \left[\overline{Q}_{ij} \right]_k [\alpha_i] [\beta_j] \Delta T(z, t) \Delta m(z, t) \mathrm{d}z$$

Equation (6.15) is the most general formulation and it is seen that without the surface load term there are 30 terms to represent the strain energy in the composite material panel. Referring to Equation (2.62) and the ensuing discussion it is seen that if the laminate has no stretching-shearing coupling ($A_{16} = A_{26} = 0$) then two terms would be dropped; if no twisting stretching coupling ($B_{16} = B_{26} = 0$), two more would be dropped; likewise two more dropped if there were no bending-twisting coupling ($D_{16} = D_{26} = 0$). If the laminate were symmetric about the mid-plane, a very common construction, then five B_{ij} terms would be cancelled out, because of no bending-stretching coupling.

6.4.2. Moisture and Temperature Distributions

Having qualitatively discussed the hygrothermal effects earlier; it is important to be able to quantitatively determine $\Delta T(z)$ and $\Delta m(z)$ given a set of surface conditions. Of course these quantities could also be a function of x and y as well but for many applications with thin walled structure such as plates and shells the most important direction for heat or moisture flux is through the thickness.

Pipes, Vinson and Chou [4] formulated the moisture concentration thickness distribution for thin walled beam, plate and shell structures using the classical diffusion equation,

$$D_z \frac{\partial^2 (\Delta m)}{\partial z^2} - \frac{\partial (\Delta m)}{\partial t} = 0 \tag{6.16}$$

where D_z is the moisture diffusion constant for the composite in the thickness direction and t is time.

For the case that the thin walled structure is instantaneously exposed to a moisture concentration Δm_0, at time $t = 0$, on both the top and bottom surfaces, the transient solution of (6.16) is the following solution:

$$\Delta m(z, t) = \Delta m_0 + \sum_{n=0}^{\infty} m_n \cos a_n z \tag{6.17}$$

where

$$a_n = \frac{(2n + 1)\pi}{h} \tag{6.18}$$

and

$$m_n = \frac{4}{\pi} \left\{ \frac{(-1)^n}{2n + 1} \, e^{-a_n^2 D_z t} \right\} \tag{6.19}$$

If only the upper surface of the thin walled structure is suddenly exposed to a moisture concentration Δm_0 at $t = 0$, the transient solution is:

$$\Delta m(z, t) = \Delta m_0 \left\{ 1 - \sum_{n=0}^{\infty} P_n \cos q_n (z + h/2) \right\} \tag{6.20}$$

where

$$q_n = \frac{(2n + 1)\pi}{2h} \tag{6.21}$$

and

$$P_n = \frac{4}{\pi} \left\{ \frac{(-1)^n}{(2n + 1)} \, e^{-q_n^2 D_z t} \right\} \tag{6.22}$$

The solutions, (6.17) and (6.20) are also valid for transient thermal distributions if Δm is replaced by ΔT and D_z becomes the thermal diffusion.

For a steady state case the transient solutions die out and can be represented far more easily by simple linear functions in the z-direction, such as

$$\Delta m(z) = \frac{1}{2}\left[\Delta m\left(\frac{h}{2}\right) - \Delta m\left(-\frac{h}{2}\right)\right] + \frac{z}{h}\left[\Delta m\left(\frac{h}{2}\right) - \Delta m\left(-\frac{h}{2}\right)\right] \quad (6.23)$$

and there is obviously an analogous expression for $\Delta T(z)$.

From these solutions all of the thermal and moisture stress resultants and stress couples can be formulated straightforwardly.

Obviously if the matrix of the composite structure is a metal or ceramic, the Δm, hygrothermal terms can be neglected; they only pertain to polymeric matrix composites.

6.3.3. Boundary Conditions

Solutions are found for three types of boundary conditions; all edges simply supported, all edges clamped, and two opposite edges simply supported, the other two clamped. Because of the necessity to include asymmetric lamina stacking sequences, resulting in bending-stretching coupling, these boundary conditions must be more specifically specified. Following Whitney, and Wu and Vinson [5] two simply supported and two clamped boundary conditions are defined as:

$$
\begin{aligned}
&S1: w_n = M_n = u_{0n} = N_{nt} = 0 \\
&S2: w_n = M_n = N_n = u_{0t} = 0 \\
&C1: w_n = \bar{\alpha}_n = u_{0n} = N_{nt} = 0 \\
&C2: w_n = \bar{\alpha}_n = N_n = u_{0t} = 0
\end{aligned}
\quad (6.24)
$$

In the above, w is the lateral deflection, u_0 is the in-plane mid-surface displacement, M is the stress couple, N is the in-plane stress resultant, $\bar{\alpha}$ is the rotation, and n is the direction normal to the edge, while t is the direction tangent to the edge.

6.4.4. Solution Approach

In this example the Rayleigh-Ritz method is used. Earlier, for the beam, no assumptions were made regarding the form of the displacements for the solution, hence an Euler-Lagrange differential equation (6.13) resulted, which subsequently needs to be solved. However, the

Rayleigh-Ritz procedure includes specifying a "guessed at" form of the displacement solution. In this case the resulting Euler-Lagrange equations are algebraic, and if good assumed displacements are used, provide very accurate solutions.

Good assumed displacements are more formally termed admissible functions, defined as those which at least satisfy all geometric boundary conditions, that is, displacement constraints and rotation constraints. Better yet are those functions which, in addition, satisfy part or all of the stress boundary conditions (that is, stress couples, in-plane and shear resultants).

In studies by Flaggs [6], Smith [7] and Linsenmann [8] which included transverse shear deformations, the rotations $\bar{\alpha}$ and $\bar{\beta}$ were determined as functions of the lateral deflection, $w(x, y)$ through analyzing a beam in bending with transverse shear deformation effects included. However, here a functional form for these variables will be assumed following Wu and Vinson [5].

For the boundary conditions $S1$,

$$w(x, y) = \sum_{m_{\text{odd}}}^{\infty} \sum_{n_{\text{odd}}}^{\infty} W_{mn} \sin \lambda_m x \sin \lambda_n y \tag{6.25}$$

$$\bar{\alpha}(x, y) = \sum_{m_{\text{odd}}}^{\infty} \sum_{n_{\text{odd}}}^{\infty} \Gamma_{mn} \sin \lambda_m x \cos \lambda_n y \tag{6.26}$$

$$\bar{\beta}(x, y) = \sum_{m_{\text{odd}}}^{\infty} \sum_{n_{\text{odd}}}^{\infty} \Lambda_{mn} \cos \lambda_m x \sin \lambda_n y \tag{6.27}$$

$$u_0(x, y) = \sum_{m_{\text{odd}}}^{\infty} \sum_{n_{\text{odd}}}^{\infty} U_{mn} \sin \lambda_m x \cos \lambda_n y \tag{6.28}$$

$$v_0(x, y) = \sum_{m_{\text{odd}}}^{\infty} \sum_{n_{\text{odd}}}^{\infty} V_{mn} \cos \lambda_m x \sin \lambda_n y \tag{6.29}$$

The form for the $S2$ boundary condition will be identical to the above for the first three variables. The last two variables become for $S2$:

$$u_0(x, y) = \sum_{m_{\text{odd}}}^{\infty} \sum_{n_{\text{odd}}}^{\infty} U_{mn} \cos \lambda_m x \sin \lambda_n y \tag{6.30}$$

$$v_0(x, y) = \sum_{m_{\text{odd}}}^{\infty} \sum_{n_{\text{odd}}}^{\infty} V_{mn} \sin \lambda_m x \cos \lambda_n y \tag{6.31}$$

The assumed forms of displacements and rotations for the $C1$ clamped boundary conditions are the characteristic beam functions originally formulated by Warburton [9], extended by Wu and Vinson [5], and completely characterized by Young and Felgar [10,11]. They are:

$$w(x, y) = \sum_{m=\text{odd}}^{\infty} \sum_{n_{\text{odd}}}^{\infty} W_{mn}\phi_{\omega_m}(x)\phi_{\omega_n}(y) \tag{6.32}$$

$$\bar{\alpha}(x, y) = \sum_{m_{\text{odd}}}^{\infty} \sum_{n_{\text{odd}}}^{\infty} \Gamma_{mn}\phi_{\bar{\alpha}m}(x)\phi_{\bar{\alpha}n}(x, y) \tag{6.33}$$

$$\bar{\beta}(x, y) = \sum_{m_{\text{odd}}}^{\infty} \sum_{n_{\text{odd}}}^{\infty} \Lambda\phi_{\bar{\beta}m}(x)\phi_{\bar{\beta}m}(x)\phi_{\bar{\beta}n}(y) \tag{6.34}$$

$$u_0(x, y) = \sum_{m_{\text{odd}}}^{\infty} \sum_{n_{\text{odd}}}^{\infty} U_{mn} \sin \lambda_m x \cos \lambda_n y \tag{6.35}$$

$$v_0(x, y) = \sum_{m_{\text{odd}}}^{\infty} \sum_{n_{\text{odd}}}^{\infty} V_{mn} \cos \lambda_m x \sin \lambda_n y \tag{6.36}$$

where

$$\left.\begin{aligned}
\phi_{\omega m}(x) &= \cos \mu_m\left(\frac{x}{a} - \frac{1}{2}\right) + \eta_m \cosh \mu_m\left(\frac{x}{a} - \frac{1}{2}\right) \\
\phi_{\bar{\alpha}m}(x) &= \sin \mu_m\left(\frac{x}{a} - \frac{1}{2}\right) - \eta_m \sinh \mu_m\left(\frac{x}{a} - \frac{1}{2}\right) \\
\phi_{\bar{\beta}m}(x) &= \cos \mu_m\left(\frac{x}{a} - \frac{1}{2}\right) + \eta_m \cosh \mu_m\left(\frac{x}{a} - \frac{1}{2}\right)
\end{aligned}\right\} \tag{6.37}$$

and

$$\eta_m = \sin(\mu_m/2)/\sinh(\mu_m/2) \tag{6.38}$$

and

$$\tan(\mu_m/2) + \tanh(\mu_m/2) = 0 \tag{6.39}$$

By interchanging $\bar{\alpha}$ and $\bar{\beta}$, substituting y for x, n for m, and b for a, (6.37) yields the proper functions for the y-direction in (6.32) through (6.36). The solution of the transcendental equation (6.39) provides the necessary values of μ_m and μ_n in (6.38) and (6.37). The solutions are

$$\mu_1 = 1.50562\pi$$

$$\mu_2 = 2.49975\pi, \quad \mu_k = \left(k + \tfrac{1}{2}\right)\pi \quad \text{for} \quad k \geqslant 3 \tag{6.40}$$

In the $C2$ boundary conditions u_0 and v_0 are given by (6.30) and (6.31).

For two edges simply supported, and the two opposite edges clamped, the following forms are employed where $x = 0$ and a are the clamped boundary conditions and $y = 0$ and b are the simply supported edges,

$$w(x, y) = \sum_{m_{\text{odd}}}^{\infty} \sum_{n_{\text{odd}}}^{\infty} w_{mn}\phi_{\omega m} \sin \lambda_n y \tag{6.41}$$

$$\bar{\alpha}(x, y) = \sum_{m_{\text{odd}}}^{\infty} \sum_{n_{\text{odd}}}^{\infty} \Gamma_{mn}\phi_{\bar{\alpha} m} \cos \lambda_n y \tag{6.42}$$

$$\bar{\beta}(x, y) = \sum_{m_{\text{odd}}}^{\infty} \sum_{n_{\text{odd}}}^{\infty} \Lambda_{mn}\phi_{\bar{\beta} m} \sin \lambda_n y \tag{6.43}$$

where

$$\lambda_n = n\pi/b$$

The $u_0(x, y)$ and $v_0(x, y)$ variables are given by either (6.28) and (6.29) or (6.30) and (6.31) depending upon the desired boundary condition.

6.4.5. Calculation of Deflections, Rotations and Stresses

The values of the coefficients W_{mn}, Γ_{mn}, Λ_{mn}, U_{mn}, and V_{mn} are obtained from substituting the above equations into Equation (6.15), taking variations, performing integrations, and solving the set of simultaneous equations. Following that, the above coefficients are substituted into (6.25) through (6.43) to find the displacements and rotations. From these the laminate strains are found from Equation (2.42). Then the stresses everywhere in each lamina are calculated using Equation (2.38), after determining in the thermal and hygroscopic contributions.

6.4.6. Numerical Examples

The following examples indicate the importance of hygrothermal effects on polymeric matrix composite materials. To efficiently effect the

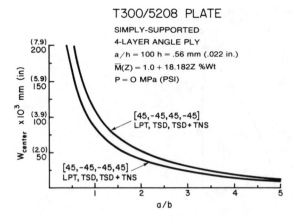

Fig. 6.2.

calculations, a computer program was developed, and is described by Sloan and Vinson [12]. The program analyzes rectangular laminated plates, symmetric or asymmetric with respect to the plate mid-plane, including clamped, simply supported or combinations of the two boundary conditions. The loads included are through the thickness moisture and temperature gradients and uniform lateral pressures. The output gives deflections, rotations, stress resultants, stress couples, strains and stresses in the plate. Options are included to perform analyses based on linear plate theory (LPT), plate theory including transverse shear deformation (TSD) and plate theory including both transverse shear deformation and transverse normal strains (TSD + TNS).

In the numerical examples shown herein, odd terms up through $m = n = 7$ were used. With this the deflections for the simply supported plate are within 0.1% of the exact solution, and 0.7% for the clamped plate case.

The material used is $T300/5208$ graphite epoxy. Properties for 21°C (70°F) and 177°C (350°F) were used with a linear interpolation used for properties at intermediate temperatures. The properties used are given in reference [12].

Figure 6.2 shows the center deflection of a four layer symmetric and asymmetric angle ply, for the case of simply supported laminates using the various plate theories for various a/b ratios. It is seen that for $a/h = 100$ and variable a/b ratios the effect of both TSD and TSD + TNS on the center deflection is small. Each lamina is 0.14 mm (0.0055 in.). It is noted that the center deflection of the asymmetric laminate is greater than that of the symmetric laminate as would be expected.

Figure 6.3 shows the moisture distribution for single surface absorp-

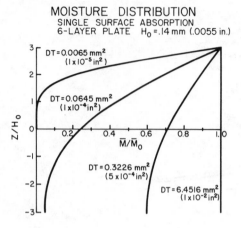

MOISTURE DISTRIBUTION
SINGLE SURFACE ABSORPTION
6-LAYER PLATE H_0 = .14 mm (.0055 in.)

Fig. 6.3.

tion at the time intervals t (where D is the diffusion coefficient) for a six layer laminate.

Consider a $[0, 45, -45]_{2S}$ symmetric plate with clamped boundary conditions and unrestrained in-plane. The loads considered are a lateral pressure of 0.014 MPa (2 psi) on the upper surface, a uniform temperature of 66°C (150°F) and a sudden moisture change to $M_0 = 1.0\%$ on the top surface at $Dt = 0$. Figures 6.4 – 6.6 show the stresses in the x, y and xy directions at various times. It is seen that S_x and S_y on the top, $\theta = 0°$, layer are compressive as expected. The maximum S_x compressive stress of approximately 62 MPa (9 ksi) occurs in the fully saturated condition, while the maximum S_y stress of 69 MPa (10 ksi) occurs at early times.

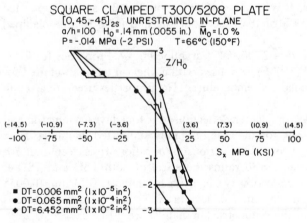

SQUARE CLAMPED T300/5208 PLATE
$[0, 45, -45]_{2S}$ UNRESTRAINED IN-PLANE
a/h=100 H_0 = .14 mm (.0055 in.) $\overline{M}_0$ = 1.0 %
P = -.014 MPa (-2 PSI)   T = 66°C (150°F)

Fig. 6.4.

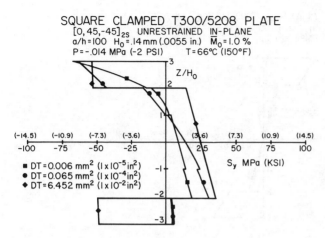

Fig. 6.5.

Also, of interest is the development of compressive stresses in the bottom 0° layer in the saturated condition. The maximum tensile stress of approximately 34.5 MPa (5 ksi) occurs at late times in the bottom, $\theta = \pm 45°$ layer. The shear stress, S_{xy}, increases with increased moisture concentration. The maximum shear stress is approximately 86 MPa (12.5 ksi) with large shear stress gradients between the $\pm 45°$ layers. These large stress gradients imply high interlaminar shear stresses as discussed by Pipes, Vinson and Chou [4].

To show the effect of a hygrothermal environment, Figure (6.7) shows S_x, S_y and S_{xy} stress distributions for the same plate with all the loads except moisture.

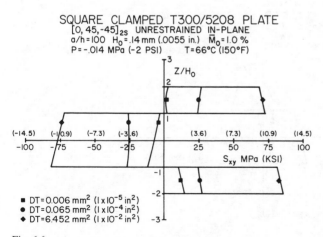

Fig. 6.6.

188

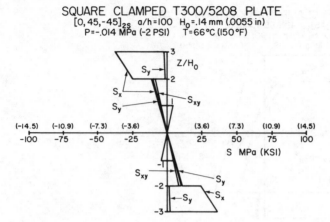

SQUARE CLAMPED T300/5208 PLATE
[0, 45, −45]$_{2s}$ a/h=100 H$_0$=.14 mm (.0055 in)
P=−.014 MPa (−2 PSI)   T=66°C (150°F)

Fig. 6.7.

It is seen that the hygrothermal environment causes increases in S_y in the 0° layers and large increases in the shear stresses, S_{xy}. Figure 6.8 shows S_{xy} in the same plate as Figure 6.6 but with in-plane expansions restrained.

It is seen that the shear stresses have markedly decreased in the middle layers with a maximum stress of only 34.5 MPa (5 ksi), a value of less than half of the unrestrained case. This would indicate a decided advantage to designing panels in a hygrothermal environment with in-plane restraints to reduce in-plane shear stresses without significantly increasing the S_x and S_y stresses. It also reduces the possibility of delamination caused by high interlaminar shear stresses.

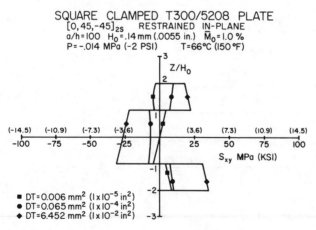

SQUARE CLAMPED T300/5208 PLATE
[0,45,−45]$_{2S}$ RESTRAINED IN-PLANE
a/h=100 H$_0$=.14 mm (.0055 in.) $\bar{M}_0$=1.0 %
P=−.014 MPa (−2 PSI)   T=66°C (150°F)

■ DT=0.006 mm^2 (1×10^{-5} in^2)
● DT=0.065 mm^2 (1×10^{-4} in^2)
◆ DT=6.452 mm^2 (1×10^{-2} in^2)

Fig. 6.8.

It is also concluded in the unabridged study [12] that it is important to include transverse shear deformation as well as hygrothermal effects in any plate analysis, but that transverse normal deformations have no appreciable influence.

6.5. Elastic Stability of Generally Laminated Composite Panels Including Hygrothermal Effects

Not only is it necessary to insure that a panel is not overstressed when subjected to lateral and in-plane loads, but it must also not buckle when subjected to in-plane compressive loads. In fact, the hygrothermal effects can be most deleterious because if a thin walled structure is subjected to heat and moisture on one side, even a laminate originally designed to be symmetric becomes asymmetric. In this section a general buckling theory is formulated which accounts for the hygrothermal effects as well as transverse shear deformation, and all of the couplings discussed in earlier sections, which was performed by Flaggs [6].

Again the Theorem of Minimum Potential Energy is used, so the strain energy of the plate is identical to that previously used, namely Equation (6.15). However, in the absence of a lateral load $p(x, y)$, the last term of Equation (6.15) is absent, but in its place are the effects of in-plane stress resultants N_x, N_{xy} and N_y, which can cause an elastic instability. This problem has been studied by Flaggs [6]. Hence, Equation (6.15) becomes

$$V = \int_R W \, dR \, (\text{See Eq. } 6.15) - \iint \left\{ N_x \left[\frac{\partial u_0}{\partial x} + \frac{1}{2} \left(\frac{\partial w}{\partial x} \right)^2 \right] \right.$$

$$\left. + N_{xy} \left[\frac{\partial u_0}{\partial y} + \frac{\partial v_0}{\partial x} + \frac{\partial w}{\partial x} \frac{\partial w}{\partial y} \right] + N_y \left[\frac{\partial v_0}{\partial y} + \frac{1}{2} \left(\frac{\partial w}{\partial y} \right)^2 \right] \right\} dA \qquad (6.41)$$

These terms are treated in detail in References 1 and 2, and will not be developed in detail here – but they are standard for an isotropic panel as well as a composite laminate.

The buckling loads, N_x, N_{xy} or N_y, are determined by finding the value of the load at which bifurcation occurs, that is, loads at which the plate can be in equilibrium in both a straight configuration (that is, $w = 0$) and in a slightly deformed ($w \neq 0$) configuration. This is accomplished through setting the variation of V in (6.41) equal to zero, as (6.2). This operation results in an eigenvalue problem which can be solved for non-trivial solutions that are discrete values of the applied loads. The lowest critical load is the actual physical buckling load. Unlike solving for several natural frequencies, all of which could be important, only the lowest buckling load has any physical meaning.

In the last section, u_0, v_0, w, $\bar{\alpha}$ and $\bar{\beta}$ were all considered to be unknowns to be found through the solution. In this study, simply to illustrate another approach, $\bar{\alpha}$ and $\bar{\beta}$, the rotations, are solved for in terms of the lateral deflection w. To do this, consider the laminate to be a beam, hence changing (6.15) to only have an x-dependence. Solving that problem results in three equations:

$$A_{11}\frac{d^2 u_0}{dx^2} + B_{11}\frac{d^2\bar{\alpha}}{dx^2} = 0 \tag{6.42}$$

$$A_{55}\frac{d\bar{\alpha}}{dx} - A_{55}\frac{d^2 w}{dx^2} - N_x\frac{d^2 w}{dx^2} = 0 \tag{6.43}$$

$$-A_{55}\bar{\alpha} - A_{55}\frac{dw}{dx} + B_{11}\frac{d^2 u}{dx^2} + D_{11}\frac{d^2\bar{\alpha}}{dx^2} = 0 \tag{6.44}$$

These equations can be solved for $\bar{\alpha}$, the result being as follows, allowing for a y-dependence, too,

$$\bar{\alpha}(x, y) = -\frac{\partial w}{\partial x} - \left(\frac{D_{11}A_{11} - B_{11}^2}{A_{11}}\right)\left(\frac{A_{55} - N_x}{A_{55}^2}\right)\frac{\partial^3 w}{\partial x^3} \tag{6.45}$$

Similarly, assuming a beam in the y-direction, it is found that

$$\bar{\beta}(y, x) = -\frac{\partial w}{\partial y} - \left(\frac{D_{22}A_{22} - B_{22}^2}{A_{22}}\right)\left(\frac{A_{44} - N_y}{A_{44}^2}\right)\frac{\partial^3 w}{\partial y^3} \tag{6.46}$$

(6.45) and (6.46) can now be substituted into (6.41) so that the potential energy expression contains only u_0, v_0 and w as the unknowns.

As a good set of examples to study to investigate various effects, from Equation (6.24) the simply supported case $S1$ and the clamped case $C1$ are chosen. Hence, the forms of displacements chosen are given by (6.25), (6.28) and (6.29) for the $S1$ case and for the $C1$ case (6.35) and (6.36) are chosen but Flaggs and Vinson chose a form for w, differing from (6.32). It must be remembered that an admissible function satisfies at least the geometric boundary conditions but that function is not unique, hence in this case the following is used

$$w(x, y) = \sum_{m=1}^{\infty}\sum_{n=1}^{\infty} W_{mn}\left[\cos\frac{(m-1)\pi x}{a} - \cos\frac{(m+1)\pi x}{a}\right]$$

$$\times\left[\cos\frac{(n-1)\pi y}{b} - \cos\frac{(n+1)\pi y}{b}\right] \tag{6.47}$$

These displacements can now be substituted into (6.15). Then taking the variation with respect to the unknown amplitudes W_{mn}, Γ_{mn} and Λ_{mn} results in the eigenvalue problem below, here we simply set $N_y = \lambda N_x$;

$$
\begin{bmatrix} E_{11} & E_{12} & E_{13} \\ E_{21} & E_{22} & E_{23} \\ E_{31} & E_{32} & E_{33} \end{bmatrix} \begin{Bmatrix} \Gamma_{mn} \\ \Lambda_{mn} \\ W_{mn} \end{Bmatrix} = N_x \begin{bmatrix} 0 & 0 & F_{13} \\ 0 & 0 & F_{23} \\ F_{31} & F_{32} & F_{33} \end{bmatrix} \begin{Bmatrix} \Gamma_{mn} \\ \Lambda_{mn} \\ W_{mn} \end{Bmatrix}
$$

$$
+ N_x^2 \begin{bmatrix} 0 & 0 & 0 \\ 0 & 0 & 0 \\ 0 & 0 & G_{33} \end{bmatrix} \begin{Bmatrix} \Gamma_{mn} \\ \Lambda_{mn} \\ W_{mn} \end{Bmatrix} \qquad (6.48)
$$

where the E_{ij}, F_{ij} and G_{ij} quantities are symbolic but can be found through the variational process.

(6.48) can be simplified by uncoupling the third set of simultaneous linear algebraic equations, by substituting for the Γ_{mn} and Δ_{mn} equations in terms of the W_{mn}, resulting in

$$
[E]\{W_{mn}\} = N_x[F]\{W_{mn}\} + (N_x)^2[G]\{W_{mn}\} \qquad (6.49)
$$

The solution of (6.49) for the buckling load N_x is done numerically and the computer program is available in Appendix Two of Reference 6.

With the general buckling theory developed above a parametric study is shown as follows to investigate the elastic stability of generally laminated composite plates including hygrothermal effects. The computer program used for the numerical analysis calculates the buckling loads for a thin laminated plate ($\sigma_z = e_z = 0$) without transverse shear deformation (LPT), with transverse shear deformation (TSD), and for a generally laminated plate with both transverse shear and normal deformation (3-D). The buckling load of a symmetrically and unsymmetrically laminated $T300/5208$ graphite-epoxy plate uniaxially loaded is calculated using laminated plate theory and the present analysis for simply supported boundary conditions. The lamina properties are given in Reference 6.

For a symmetrically laminated $[0, 45, -45, 90]_{4s}$ $T300/5208$ graphite epoxy composite plate with simply supported and clamped boundary conditions, the buckling loads using the generally laminated plate theory (3-D) are calculated for both steady state and transient hygrothermal conditions. It is assumed that the lamina properties vary linearly between 70° and 350°F for those which are given as temperature dependent.

Figures 6.9 and 6.10 show the effects on the applied buckling load N_x

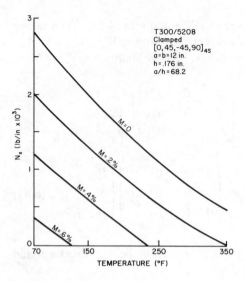

Fig. 6.9.

of different steady state hygrothermal environments for clamped and simply supported boundary conditions. The effect of single surface moisture diffusion at different elevated temperatures is shown in Figure 6.11 for a clamped laminated plate. Figure 6.12 gives the moisture distribution profiles for the different transient cases considered in Figure 6.11.

For engineering applications, the designer must know the ranges over which different theories provide valid design information. For this purpose, the buckling loads of two different laminates $[0_{32}]$ and $[0, 45, -45, 90]_{4s}$, at steady state hygrothermal conditions of 160°F and $M = 0.4\%$ are calculated using the thin laminated plate theory with and without transverse shear deformation and the general laminated plate theory with both transverse shear and normal deformation with clamped boundary conditions for different a/h and a/b aspect ratios. The graphite-epoxy material system is $GY70/339$ where the lamina properties at 160°F are found in Reference 6.

The effect of the aspect ratio (a/h) on the buckling load of a $[0_{32}]$ laminated plate is shown in Figure 6.13. For aspect ratios (a/h) of 68.2,

Table 6.1.

	CLASSICAL [19]	LPT	TSD	3-D
$[0, 45, -45, 90]_s$, N_x (lb/in)	16.546	16.545	16.542	16.542
$[45, -45, 45, -45]$, N_x (lb/in)	2.279	2.3954	2.3954	2.4013

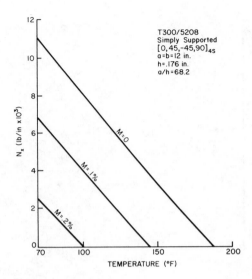

Fig. 6.10.

34.1 and 17.05, the effect of a/b from 0.25 to 2 is seen in Figures 6.14, 6.15 and 6.16 respectively. For a $[0, 45, -45, 90]_{4s}$ laminated plate, the effect of the aspect ratio (a/h) on the buckling load is given as a comparison in Figure 6.17.

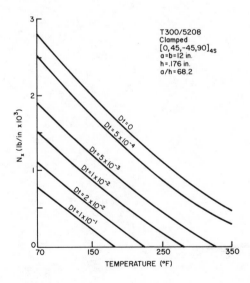

Fig. 6.11.

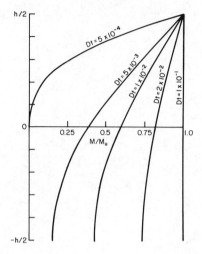

Fig. 6.12.

From the results of the parametric study, several important conclusions can be made. First, elevated hygrothermal conditions severely reduce the compressive loads necessary to buckle laminated composite plates. For the simply supported $T300/5208$ graphite-epoxy laminated

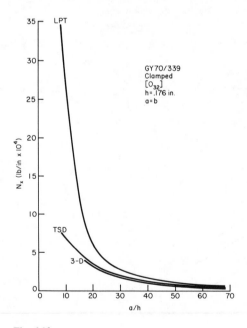

Fig. 6.13.

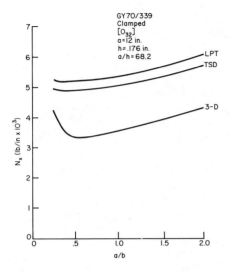

Fig. 6.14.

[0, ±45, 90]$_{4s}$ plate considered in the analysis using the general buckling theory formulated in this study, it was shown that an equilibrium moisture content of only 0.25% above the stress-free level will cause buckling at room temperatures, while at a temperature of just 145°F, buckling will occur at $M = 0.1\%$. Increasing the boundary constraints by clamping the edges increases the buckling load and reduces slightly the adverse effects of an elevated hygrothermal environment. For the same

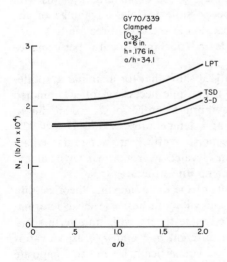

Fig. 6.15.

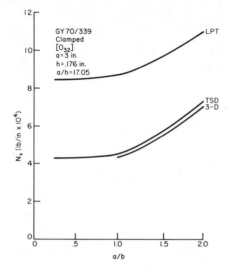

Fig. 6.16.

laminated composite plate considered above with clamped boundary conditions, an equilibrium moisture content of less than 0.7% above the stress-free level causes elastic instability at room temperature. As expected, the effects of moisture absorption increase with time, reaching their maximum value at equilibrium. Comparison of the hygrothermal stress resultants from the $T300/5208$ graphite-epoxy system ($E_1/E_2 = 13.9$) with those from the $GY70$-339 graphite-epoxy system ($E_1/E_2 = 63.6$) for the $[0, \pm45, 90]_{4s}$ laminate show that by use of material systems with high (E_1/E_2) ratios, it should be possible to take advantage of the negative thermal coefficient of expansion in the fiber direction to reduce and possibly even eliminate the degradation in structural performance due to moisture effects.

As an aside, Shen and Springer [13] show that for a similar graphite-epoxy material system, $T300/1034$, the time necessary for a laminated plate 0.176 in. thick to reach an equilibrium content of $M = 0.5\%$ due to single surface moisture diffusion at a temperature of 170°F and 90% humidity is 1.25 years. So while the moisture effects are not an immediate problem long term exposure to an elevated hygrothermal environment can have devastating effects on a composite structure.

Second, comparison of the results of the three buckling theories, thin laminated plate theory with (TSD) and without transverse shear deformation (LPT) and a general laminated plate theory with transverse shear and normal deformation (3-D), for different a/b and a/h aspect ratios summarized in Tables 6.2 and 6.3 for symmetrically laminated composite plates of $GY70/339$ at $T = 160°F$, $M = 0.4\%$ shows that inclusion of

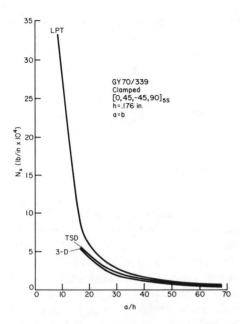

Fig. 6.17.

transverse shear deformation is necessary for the analysis of a laminated composite fabricated from a highly orthotropic material with clamped boundary conditions under elevated hygrothermal conditions.

For thicker plates ($a/h \leqslant 34.1$), neglecting transverse shear deformation results is grossly overestimating the buckling load. Comparison between the theory with just transverse shear deformation (TSD) and the general theory with both transverse shear and normal deformation for symmetrically laminated plates with hygrothermal effects. For thin plates ($a/h = 68.2$), additional reduction in buckling loads up to 31.45% were seen while for thicker plates ($a/h \leqslant 34.1$), differences of up to 4.68% were

Table 6.2. TSD vs. LPT, Reduction (%) in N_x due to Transverse Shear Deformation; $[0_{32}]$, ($[0, 45, -45, 90]_{4s}$)

a/b a/h	2	1	0.5	0.38	0.25
68.2	5.44	5.69 (4.35)	5.89	5.78	
34.1	17.01	19.19 (14.75)	19.26	19.24	19.23
17.05	33.78	48.55 (34)	48.97	48.97	48.97
8.52		78.21			

Table 6.3. 3-D vs. TSD, Additional Reduction (:) in N_x due to General Theory; $[0_{32}]$, $([0, 45, -45, 90]_{4s})$

a/b a/h	2	1	0.5	0.33	0.25
68.2	23.17	28.21 (10.7)	31.45	18.88	13.57
34.1	4.68	2.75 (2.81)	2.01	2.06	2.08
17.05	2.01	0.88 (0.38)			

noted as the ratio a/b increased. The initially surprising large effect on thin plates is due to three things: 1) the material system is more highly orthotropic at elevated hygrothermal conditions, $E_1/E_2 = 47.6$ at 70°F vs. $E_1/E_2 = 63.6$ at 160°F; 2) hygrothermal stress resultants are the same for any plate with the same material properties, stacking sequence and thickness regardless of span dimensions; and 3) hygrothermal stress resultants are larger in the general theory formulation due to additional Poisson's effects introduced by inclusion of transverse normal deformation.

6.6. Theorem of Minimum Potential Energy for a Laminated Composite Cylindrical Shell

It is clear that the Theorem of Minimum Potential Energy has certain advantages over the solution of governing differential equation and boundary conditions for beam problems, and has many advantages for the solution of composite material plates with hygrothermal effects. However, for shell problems discretion must be exercised. For a shell subjected to a static load, to determine the stresses accurately requires choosing a set of displacement functions which fairly accurately portray the deflections in the bending boundary layer, where the maximum stresses almost always occur. That is a formidable task, and to say the least, requires intricate displacement functions and complicated integrals of those functions. Hence, the Energy Theorem approach would probably be utilized for static stress problems as a last resort.

However, for eigenvalue problems – both the determination of natural frequencies and buckling loads – the Theorem of Minimum Potential Energy provides a powerful tool, particularly with complicated geometries, stacking sequences, hygrothermal effects, and the complication of transverse shear deformation.

The Minimum Potential Energy expression is derived herein for a circular cylindrical shell of generalized stacking sequence, transverse shear deformation, and hygrothermal effects.

Again, the potential energy, V, neglecting the body force term,

$$V = \int_R W \, dR - \int_{S_T} T_i u_i \, dS \tag{6.50}$$

where

$$W = \tfrac{1}{2} \sigma_{ij} \epsilon_{ij} \tag{6.51}$$

$$W = \tfrac{1}{2} \sigma_x \bar{\epsilon}_x + \tfrac{1}{2} \sigma_\theta \bar{\epsilon}_\theta + \sigma_{x\theta} \bar{\epsilon}_{x\theta} + \sigma_{\theta r} \bar{\epsilon}_{\theta r} + \sigma_{rx} \bar{\epsilon}_{rx} \tag{6.52}$$

where bars denote isothermal strains.
Because the shell is laminated with each ply oriented differently, then for the kth ply,

$$W_k = \tfrac{1}{2} \sigma_{x_k} \bar{\epsilon}_{x_k} + \tfrac{1}{2} \sigma_{\theta_k} \bar{\epsilon}_{\theta_k} + \sigma_{x\theta_k} \bar{\epsilon}_{x\theta_k} + \sigma_{\theta z_k} \bar{\epsilon}_{\theta z_k} + \sigma_{rx_k} \bar{\epsilon}_{rx_k} \tag{6.53}$$

The constitutive equations for the kth lamina of the shell are:

$$
\begin{bmatrix} \sigma_x \\ \sigma_\theta \\ \sigma_{\theta z} \\ \theta_{xz} \\ \sigma_{x\theta} \end{bmatrix}_k =
\begin{bmatrix}
\overline{Q}_{11} & \overline{Q}_{12} & 0 & 0 & 2\overline{Q}_{16} \\
\overline{Q}_{12} & \overline{Q}_{22} & 0 & 0 & 2\overline{Q}_{26} \\
0 & 0 & 2\overline{Q}_{44} & 2\overline{Q}_{45} & 0 \\
0 & 0 & 2\overline{Q}_{45} & 2\overline{Q}_{55} & 0 \\
\overline{Q}_{16} & \overline{Q}_{26} & 0 & 0 & 2\overline{Q}_{66}
\end{bmatrix}_k
\begin{bmatrix}
\epsilon_x - \alpha_{x_k} \Delta T - \beta_{x_k} \Delta m \\
\epsilon_\theta - \alpha_{\theta_k} \Delta T - \beta_{\theta_k} \Delta m \\
\epsilon_{\theta z} \\
\epsilon_{xz} \\
\epsilon_{x\theta} - \alpha_{x\theta_k} \Delta T - \beta_{x\theta_k} \Delta m
\end{bmatrix}_k
\tag{6.54}
$$

In these constitutive relations the total strains are written as:

$$\epsilon_x = \bar{\epsilon}_{x_k} + \alpha_{x_k} \Delta T + \beta_{x_k} \Delta m$$
$$\epsilon_\theta = \bar{\epsilon}_{\theta z} + \alpha_{\theta_k} \Delta T + \beta_{\theta_k} \Delta m$$
$$\epsilon_{\theta z} = \bar{\epsilon}_{\theta z} \tag{6.55}$$
$$\epsilon_{xz} = \bar{\epsilon}_{xz}$$
$$\epsilon_{\theta x} = \bar{\epsilon}_{x\theta_k} + \alpha_{x\theta_k} \Delta T + \beta_{x\theta_k} \Delta m$$

The assumed form of displacements are given by:

$$u(x, \theta, z) = u_0(x, \theta) + z \gamma_x(x, \theta)$$
$$v(x, \theta, z) = v_0(x, \theta) + z \gamma_\theta(x, \theta) \tag{6.56}$$
$$w = w(x, \theta) \quad \text{only}$$

The strain-displacement relations for the laminated shell are:

$$\epsilon_x = \frac{\partial u}{\partial x}$$

$$\epsilon_\theta = \frac{1}{R}\frac{\partial v}{\partial \theta} + \frac{w}{R}$$

$$\epsilon_{\theta z} = \frac{1}{2}\left(\frac{\partial v}{\partial z} + \frac{1}{R}\frac{\partial w}{\partial \theta}\right) \qquad (6.57)$$

$$\epsilon_{xz} = \frac{1}{2}\left(\frac{\partial u}{\partial z} + \frac{\partial w}{\partial x}\right)$$

$$\epsilon_{\theta x} = \frac{1}{2}\left(\frac{1}{R}\frac{\partial u}{\partial \theta} + \frac{\partial v}{\partial x}\right)$$

substituting (6.56) into (6.57) results in strain-displacement relations through the shell wall thickness.

$$\epsilon_x = \frac{\partial u_0}{\partial x} + z\frac{\partial \gamma_x}{\partial x}$$

$$\epsilon_\theta = \frac{1}{R}\frac{\partial v_0}{\partial \theta} + \frac{z}{R}\frac{\partial \gamma_\theta}{\partial \theta} + \frac{w}{R}$$

$$\epsilon_{\theta z} = \frac{1}{2}\left(\gamma_\theta + \frac{1}{R}\frac{\partial w}{\partial \theta}\right) \qquad (6.58)$$

$$\epsilon_{xz} = \frac{1}{2}\left(\gamma_x + \frac{\partial w}{\partial x}\right)$$

$$\epsilon_{x\theta} = \frac{1}{2}\left(\frac{1}{R}\frac{\partial u_0}{\partial \theta} + \frac{z}{R}\frac{\partial \gamma_x}{\partial \theta} + \frac{\partial v_0}{\partial x} + z\frac{\partial \gamma_\theta}{\partial x}\right)$$

Rewriting (6.55) one obtains,

$$\bar{\epsilon}_{x_k} = \epsilon_x - \alpha_{x_k}\Delta T - \beta_{x_k}\Delta m$$

$$\bar{\epsilon}_{\theta_k} = \epsilon_\theta - \alpha_{\theta_k}\Delta T - \beta_{\theta_k}\Delta m$$

$$\bar{\epsilon}_{\theta z_k} = \epsilon_{\theta z} \qquad (6.59)$$

$$\bar{\epsilon}_{xz_k} = \epsilon_{xz}$$

$$\bar{\epsilon}_{x\theta_k} = \epsilon_{x\theta} - \alpha_{x\theta_k}\Delta T - \beta_{x\theta_k}\Delta m$$

Rewriting (6.54) one calculates the following:

$$\sigma_{x_k} = \overline{Q}_{11_k}\left[\epsilon_x - \alpha_{x_k}\Delta T - \beta_{x_k}\Delta m\right] + \overline{Q}_{12_k}\left[\epsilon_\theta - \alpha_{\theta_k}\Delta T - \beta_{\theta_k}\Delta m\right]$$
$$+ 2\overline{Q}_{16_k}\left[\epsilon_{x\theta} - \alpha_{x\theta_k}\Delta T - \beta_{x\theta_k}\Delta m\right]$$

$$\sigma_{\theta k} = \overline{Q}_{12_k}\left[\epsilon_x - \alpha_{x_k}\Delta T - \beta_{x_k}\Delta m\right] + \overline{Q}_{22_k}\left[\epsilon_\theta - \alpha_{\theta_k}\Delta T - \beta_{\theta_k}\Delta m\right]$$
$$+ 2Q_{26_k}\left[\epsilon_{x\theta} - \alpha_{x\theta_k}\Delta T - \beta_{x\theta_k}\Delta m\right]$$

$$\sigma_{\theta z_k} = 2\overline{Q}_{44}\epsilon_{x\theta} + 2\overline{Q}_{45}\epsilon_{xz}$$

$$\sigma_{xz_k} = 2\overline{Q}_{45}\epsilon_{\theta z} + 2\overline{Q}_{55k}\epsilon_{xz}$$

$$\sigma_{x\theta_k} = \overline{Q}_{16_k}\left[\epsilon_x - \alpha_{x_k}\Delta T - \beta_{x_k}\Delta m\right] + \overline{Q}_{26_k}\left[\epsilon_\theta - \alpha_{\theta_k}\Delta T - \beta_{\theta_k}\Delta m\right]$$
$$+ 2\overline{Q}_{66_k}\left[\epsilon_{x\theta} - \alpha_{x\theta_k}\Delta T - \beta_{x\theta_k}\Delta m\right] \tag{6.60}$$

Therefore, (6.60) and (6.59) substituted into (6.53) results in

$$W_k = \tfrac{1}{2}\overline{Q}_{11_k}\epsilon_{x0}^2 + \overline{Q}_{11_k}\epsilon_{x_0}\kappa_x z + \tfrac{1}{2}\overline{Q}_{11_k}\kappa_x^2 z^2$$
$$+ \tfrac{1}{2}\overline{Q}_{22_k}\epsilon_{\theta_0}^2 + \overline{Q}_{22_k}\epsilon_{\theta_0}\kappa_\theta z + \tfrac{1}{2}\overline{Q}_{22_k}\kappa_\theta^2 z^2$$
$$- \epsilon_{x_0}\left\{\overline{Q}_{11_k}\alpha_{x_k}\Delta T + \overline{Q}_{11_k}\beta_{x_k}\Delta m + \overline{Q}_{12_k}\alpha_{\theta_k}\Delta T + \overline{Q}_{12_k}\beta_{\theta_k}\Delta m\right.$$
$$\left. + 2\overline{Q}_{16_k}\alpha_{x\theta_k}\Delta T + 2\overline{Q}_{16_k}\beta_{x\theta_k}\Delta m\right\}$$
$$- \kappa_x\left\{\overline{Q}_{11_k}z\alpha_{x_k}\Delta T + \overline{Q}_{11_k}z\beta_{x_k}\Delta m + \overline{Q}_{12_k}z\alpha_{\theta_k}\Delta T + \overline{Q}_{12_k}z\beta_{\theta_k}\Delta m\right.$$
$$\left. + 2\overline{Q}_{16_k}z\alpha_{x\theta_k}\Delta T + 2\overline{Q}_{16_k}z\beta_{x\theta_k}\Delta m\right\}$$

$+$ terms with no variation in them

$$- \epsilon_{\theta_0}\left\{\overline{Q}_{12_k}\alpha_{x_k}\Delta T + \overline{Q}_{12}\beta_{x_k}\Delta m + \overline{Q}_{22_k}\alpha_{\theta_k}\Delta T + \overline{Q}_{22_k}\beta_{\theta_k}\Delta m\right.$$
$$\left. + 2\overline{Q}_{26_k}\alpha_{x\theta_k}\Delta T + 2\overline{Q}_{26_k}\beta_{x\theta_k}\Delta m\right\}$$
$$- \kappa_\theta\left\{\overline{Q}_{12_k}\alpha_{xk}z\Delta T + \overline{Q}_{12_k}\beta_{x_k}z\Delta m + \overline{Q}_{22_k}\alpha_{\theta_k}z\Delta T + \overline{Q}_{22}\beta_{\theta_k}z\Delta m\right.$$
$$\left. + 2\overline{Q}_{26_k}\alpha_{xa_k}z\Delta T + 2\overline{Q}_{26_k}\beta_{x\theta_k}z\Delta m\right\}$$
$$- 2\epsilon_{x\theta_0}\left\{\overline{Q}_{16_k}\alpha_{x_k}\Delta T + \overline{Q}_{16_k}\beta_{x_k}\Delta m + \overline{Q}_{26_k}\alpha_{\theta_k}\Delta T + \overline{Q}_{26_k}\beta_{\theta_k}\Delta m\right.$$
$$\left. + 2\overline{Q}_{66_k}\alpha_{x\theta_k}\Delta T + 2\overline{Q}_{66_k}\beta_{x\theta_k}\Delta m\right\}$$
$$- 2\kappa_{x\theta}\left\{\overline{Q}_{16_k}\alpha_{x_k}z\Delta T + \overline{Q}_{16_k}\beta_{x_k}z\Delta m + \overline{Q}_{26_k}\alpha_{\theta_k}z\Delta T + \overline{Q}_{26_k}\beta_{\theta_k}z\Delta m\right.$$

$$+2\overline{Q}_{66}\alpha_{x\theta_k}z\Delta T + 2\overline{Q}_{66}\beta_{x\theta_k}z\Delta m\}$$

$$+2\overline{Q}_{66_k}\epsilon^2_{x\theta_0} + 4\overline{Q}_{66_k}\epsilon_{x\theta_0}\kappa_{x\theta}z + 2\overline{Q}_{66_k}\kappa^2_{x\theta}z^2 \tag{6.61}$$

where

$$\epsilon_{x_0} = \frac{\partial u_0}{\partial x}, \quad \epsilon_{\theta_0} = \frac{1}{R}\frac{\partial v_0}{\partial \theta} + \frac{w}{R}, \quad \epsilon_{x\theta_0} = \frac{1}{2}\left(\frac{1}{R}\frac{\partial u_0}{\partial \theta} + \frac{\partial v_0}{\partial x}\right)$$

$$\kappa_x = \frac{\partial \gamma_x}{\partial x}, \quad \kappa_\theta = \frac{1}{R}\frac{\partial \gamma_\theta}{\partial \theta}, \quad \kappa_{x\theta} = \frac{1}{2}\left(\frac{1}{R}\frac{\partial \gamma_x}{\partial \theta} + \frac{\partial \gamma_\theta}{\partial x}\right)$$

The strain energy for the the kth lamina of the shell is written as:

$$U_k = \int_{h_{k-1}}^{h_k}\int_0^{2\pi R}\int_0^L W_k\mathrm{d}x\mathrm{d}\zeta R\mathrm{d}\theta \tag{6.62}$$

or

$$U_k = \int_0^{2\pi R}\int_0^L\left\{\frac{1}{2}\epsilon^2_{x0}\overline{Q}_{11_k}(h_k - h_{k-1}) + \epsilon_{x_0}\kappa_x\frac{\overline{Q}_{11_k}}{2}\left(h_k^2 - h_{k-1}^2\right)\right.$$

$$+ \frac{1}{2}\kappa_x^2\frac{\overline{Q}_{11_k}}{3}\left(h_k^3 - h_{k-1}^3\right) + \frac{1}{2}\epsilon^2_{\theta_0}\overline{Q}_{22_k}(h_k - h_{k-1})$$

$$+ \epsilon_{\theta_0}\kappa_\theta\overline{Q}_{22_k}\frac{\left(h_k^2 - h_{k-1}^2\right)}{2} + \left.\frac{1}{2}\kappa_\theta^2\frac{\overline{Q}_{22_k}}{3}\left(h_k^3 - h_{k-1}^3\right)\right\}$$

$$- \epsilon_{x_0}\left[\overline{Q}_{11_k}\alpha_{x_k}\int_{h_{k-1}}^{h_k}\Delta T\mathrm{d}z + \overline{Q}_{11_k}\beta_{x_k}\int_{h_{k-1}}^{h_k}\Delta m\mathrm{d}z\right.$$

$$+ \overline{Q}_{12_k}\alpha_{\theta_k}\int_{h_{k-1}}^{h_k}\Delta T\mathrm{d}z + \overline{Q}_{12_k}\beta_{\theta_k}\int_{h_{k-1}}^{h_k}\Delta m\mathrm{d}z$$

$$+ 2\overline{Q}_{16_k}\alpha_{x\theta_k}\int_{h_{k-1}}^{h_k}\Delta T\mathrm{d}z + 2\overline{Q}_{16_k}\beta_{x\theta_k}\int_{h_{k-1}}^{h_k}\Delta m\mathrm{d}z\Bigg]$$

$$- \kappa_x\left[\overline{Q}_{11_k}\alpha_{x_k}\int_{h_{k-1}}^{h_k}\Delta Tz\mathrm{d}z + \overline{Q}_{11_k}\beta_{x_k}\int_{h_{k-1}}^{h_k}\Delta mz\mathrm{d}z\right.$$

$$+ \overline{Q}_{12_k}\alpha_{\theta_k}\int_{h_{k-1}}^{h_k}\Delta Tz\mathrm{d}z + \overline{Q}_{12_k}\beta_{\theta_k}\int_{h_{k-1}}^{h_k}\Delta mz\mathrm{d}z$$

$$+ 2\overline{Q}_{16_k}\alpha_{x\theta_k}\int_{h_{k-1}}^{h_k}\Delta Tz\mathrm{d}z + 2\overline{Q}_{16_k}\beta_{x\theta_k}\int_{h_{k-1}}^{h_k}\Delta mz\mathrm{d}z\Bigg]$$

$$-\epsilon_{\theta_0}\left[\overline{Q}_{12_k}\alpha_{x_k}\int_{h_{k-1}}^{h_k}\Delta T\mathrm{d}z+\overline{Q}_{12_k}\beta_{x_k}\int_{h_{k-1}}^{h_k}\Delta m\mathrm{d}z\right.$$

$$+\overline{Q}_{22_k}\alpha_{\theta_k}\int_{h_{k-1}}^{h_k}\Delta T\mathrm{d}z+\overline{Q}_{22_k}\beta_{\theta_k}\int_{h_{k-1}}^{h_k}\Delta m\mathrm{d}z$$

$$\left.+2\overline{Q}_{26_k}\alpha_{x\theta_k}\int_{h_{k-1}}^{h_k}\Delta T\mathrm{d}z+2\overline{Q}_{26_k}\beta_{x\theta_k}\int_{h_{k-1}}^{h_k}\Delta m\mathrm{d}z\right]$$

$$-\kappa_{\theta}\left[\overline{Q}_{12_k}\alpha_{x_k}\int_{h_{k-1}}^{h_k}\Delta Tz\mathrm{d}z+\overline{Q}_{12_k}\beta_{x_k}\int_{h_{k-1}}^{h_k}\Delta mz\mathrm{d}z\right.$$

$$+\overline{Q}_{22_k}\alpha_{\theta_k}\int_{h_{k-1}}^{h_k}\Delta Tz\mathrm{d}z+\overline{Q}_{22_k}\beta_{\theta_k}\int_{h_{k-1}}^{h_k}\Delta mz\mathrm{d}z$$

$$\left.+2\overline{Q}_{26_k}\alpha_{x\theta_k}\int_{h_{k-1}}^{h_k}\Delta Tz\mathrm{d}z+2\overline{Q}_{26_k}\beta_{x\theta_k}\int_{h_{k-1}}^{h_k}\Delta mz\mathrm{d}z\right]$$

$+$ terms with no variables which will drop out $+$

$$-2\epsilon_{x\theta_0}\left[\overline{Q}_{16_k}\alpha_{x_k}\int_{h_{k-1}}^{h_k}\Delta T\mathrm{d}z+\overline{Q}_{16_k}\beta_{x_k}\int_{h_{k-1}}^{h_k}\Delta m\mathrm{d}z\right.$$

$$+\overline{Q}_{26_k}\alpha_{\theta_k}\int_{h_{k-1}}^{h_k}\Delta T\mathrm{d}z+\overline{Q}_{26_k}\beta_{\theta_k}\int_{h_{k-1}}^{h_k}\Delta m\mathrm{d}z$$

$$\left.+2\overline{Q}_{66_k}\alpha_{x\theta_k}\int_{h_{k-1}}^{h_k}\Delta T\mathrm{d}z+2\overline{Q}_{66_k}\beta_{x\theta_k}\int_{h_{k-1}}^{h_k}\Delta m\mathrm{d}z\right]$$

$$-2\kappa_{x\theta}\left[\overline{Q}_{16_k}\alpha_{x_k}\int_{h_{k-1}}^{h_k}\Delta Tz\mathrm{d}z+\overline{Q}_{16_k}\beta_{x_k}\int_{h_{k-1}}^{h_k}\Delta mz\mathrm{d}z\right.$$

$$+\overline{Q}_{26_k}\alpha_{\theta_k}\int_{h_{k-1}}^{h_k}\Delta Tz\mathrm{d}z+\overline{Q}_{26_k}\beta_{\theta_k}\int_{h_{k-1}}^{h_k}\Delta mz\mathrm{d}z$$

$$\left.+2\overline{Q}_{66_k}\alpha_{x\theta_k}\int_{h_{k-1}}^{h_k}\Delta Tz\mathrm{d}z+2\overline{Q}_{66_k}\beta_{x\theta_k}\int_{h_{k-1}}^{h_k}\Delta mz\mathrm{d}z\right]$$

$$+2\epsilon_{\theta z}^2\overline{Q}_{44_k}(h_k-h_{k-1})+4\epsilon_{xz}\epsilon_{\theta z}\overline{Q}_{45_k}(h_k-h_{k-1})$$

$$+2\epsilon_{xz}^2\overline{Q}_{55k}(h_k-h_{k-1})+2\epsilon_{x\theta_0}^2\overline{Q}_{66}(h_k-h_{k-1})$$

$$+4\epsilon_{x\theta_0}\kappa_{x\theta}\frac{\overline{Q}_{66_k}}{2}\left(h_k^2-h_{k-1}^2\right)+2\kappa_{x\theta}^2\frac{\overline{Q}_{66_k}}{3}\left(h_k^3-h_{k-1}^3\right)\bigg\}\mathrm{d}x R\mathrm{d}\theta\quad(6.63)$$

Equation (6.63) is the strain energy for the kth lamina or ply. The total strain energy for the N ply laminate, U, is given by

$$U = \sum_{k=1}^{N} U_k \tag{6.64}$$

Specifically, the strain energy is

$$U = \int_0^{2\pi R} \int_0^L \Big\{ \tfrac{1}{2} A_{11} \epsilon_{x_0}^2 + B_{11} \epsilon_{x_0} \kappa_x + \tfrac{1}{2} D_{11} \kappa_x^2 + \tfrac{1}{2} A_{22} \epsilon_{\theta_0}^2 + B_{22} \epsilon_{\theta_0} \kappa_\theta + \tfrac{1}{2} D_{22} \kappa_\theta^2$$

$$+ 2 A_{44} \epsilon_{\theta z}^2 + 4 A_{45} \epsilon_{xz} \epsilon_{\theta z} + 2 A_{55} \epsilon_{xz}^2 + 2 A_{66} \epsilon_{x\theta_0}^2 + 4 B_{66} \epsilon_{x\theta_0} \kappa_{x\theta} + 2 D_{66} \kappa_{x\theta}^2$$

$$- \epsilon_{x0} \sum_{k=1}^{N} \Big[\overline{Q}_{11_k} \alpha_{x_k} \int_{h_{k-1}}^{h_k} \Delta T dz + \overline{Q}_{11_k} \beta_{x_k} \int_{h_{k-1}}^{h_k} \Delta m dz$$

$$+ \overline{Q}_{12_k} \alpha_{\theta_k} \int_{h_{k-1}}^{h_k} \Delta T dz + \overline{Q}_{12_k} \beta_{\theta_k} \int_{h_{k-1}}^{h_k} \Delta m dz$$

$$+ 2 \overline{Q}_{16_k} \alpha_{x\theta_k} \int_{h_{k-1}}^{h_k} \Delta T dz + 2 \overline{Q}_{16_k} \beta_{x\theta_k} \int_{h_{k-1}}^{h_k} \Delta m dz \Big]$$

$$- \kappa_x \sum_{k=1}^{N} \Big[\overline{Q}_{11_k} \alpha_{x_k} \int_{h_{k-1}}^{h_k} \Delta T z dz + \overline{Q}_{11_k} \beta_{x_k} \int_{h_{k-1}}^{h_k} \Delta m z dz$$

$$+ \overline{Q}_{12_k} \alpha_{\theta_k} \int_{h_{k-1}}^{h_k} \Delta T z dz + \overline{Q}_{12_k} \beta_{\theta_k} \int_{h_{k-1}}^{h_k} \Delta m z dz$$

$$+ 2 \overline{Q}_{16_k} \alpha_{x\theta_k} \int_{h_{k-1}}^{h_k} \Delta T z dz + 2 \overline{Q}_{16_k} \beta_{x\theta_k} \int_{h_{k-1}}^{h_k} \Delta m z dz \Big]$$

$$- \epsilon_{\theta_0} \sum_{t=1}^{N} \Big[\overline{Q}_{12_k} \alpha_{x_k} \int_{h_{k-1}}^{h_k} \Delta T dz + \overline{Q}_{12_k} \beta_{x_k} \int_{h_{k-1}}^{h_k} \Delta m dz$$

$$+ \overline{Q}_{22_k} \alpha_{\theta_k} \int_{h_{k-1}}^{h_k} \Delta T dz + \overline{Q}_{22_k} \beta_{\theta_k} \int_{h_{k-1}}^{h_k} \Delta m dz$$

$$+ 2 \overline{Q}_{26_k} \alpha_{x\theta_k} \int_{h_{k-1}}^{h_k} \Delta T dz + 2 \overline{Q}_{26_k} \beta_{x\theta_k} \int_{h_{k-1}}^{h_k} \Delta m dz \Big]$$

$$- \kappa_\theta \sum_{k=1}^{N} \Big[\overline{Q}_{12_k} \alpha_{x_k} \int_{h_{k-1}}^{h_k} \Delta T z dz + \overline{Q}_{12_k} \beta_{x_k} \int_{h_{k-1}}^{h_k} \Delta m z dz$$

$$+ \overline{Q}_{22_k} \alpha_{\theta_k} \int_{h_{k-1}}^{h_k} \Delta T z dz + \overline{Q}_{22_k} \beta_{\theta_k} \int_{h_{k-1}}^{h_k} \Delta m z dz$$

$$+ 2\overline{Q}_{26_k} \alpha_{x\theta_k} \int_{h_{k-1}}^{h_k} \Delta Tz \, dz + 2\overline{Q}_{26_k} \beta_{x\theta_k} \int_{h_{k-1}}^{h_k} \Delta mz \, dz \Bigg]$$

$$- 2\epsilon_{x\theta_0} \sum_{n=1}^{N} \left[\overline{Q}_{16_k} \alpha_{x_k} \int_{h_{k-1}}^{h_k} \Delta T \, dz + \overline{Q}_{16_k} \beta_{x_k} \int_{h_{k-1}}^{h_k} \Delta m \, dz \right.$$

$$+ \overline{Q}_{26_k} \alpha_{\theta_k} \int_{h_{k-1}}^{h_k} \Delta T \, dz + \overline{Q}_{26_k} \beta_{\theta_k} \int_{h_{k-1}}^{h_k} \Delta m \, dz$$

$$+ 2\overline{Q}_{66_k} \alpha_{x\theta_k} \int_{h_{k-1}}^{h_k} \Delta T \, dz + 2\overline{Q}_{66_k} \beta_{x\theta_k} \int_{h_{k-1}}^{h_k} \Delta m \, dz \Bigg]$$

$$- 2\kappa_{x\theta} \sum_{k=1}^{N} \left[\overline{Q}_{16_k} \alpha_{x_k} \int_{h_{k-1}}^{h_k} \Delta Tz \, dz + \overline{Q}_{16_k} \beta_{x_k} \int_{h_{k-1}}^{h_k} \Delta mz \, dz \right.$$

$$+ \overline{Q}_{26_k} \alpha_{\theta_k} \int_{h_{k-1}}^{h_k} \Delta Tz \, dz + \overline{Q}_{26_k} \beta_{q_k} \int_{h_{k-1}}^{h_k} \Delta mz \, dz$$

$$+ 2\overline{Q}_{66} \alpha_{x\theta_k} \int_{h_{k-1}}^{h_k} \Delta Tz \, dz + 2\overline{Q}_{66} \beta_{x\theta_k} \int_{h_{k-1}}^{h_k} \Delta mz \, dz \Bigg] dx \, ds$$

+ other terms which have no variables　　　　　　　　　　(6.65)

$$0 \leqslant x \leqslant L$$
$$0 \leqslant s \leqslant 2\pi R$$

For a shell that does not extend a full 360°, then the limit on the ds integral must be varied accordingly.

Keep in mind that the strain energy U as denoted by Equation (6.65) allows for

- Non-axial symmetry
- Midplane asymmetry
- transverse shear deformation
- general anistropy of the laminate
- hygrothermal effects

If axial symmetry were imposed:

$$0 = \frac{\partial}{\partial \theta} = v = \epsilon_{\theta z} = \epsilon_{x\theta} = v_0 = \epsilon_{\theta z_0} = \epsilon_{x\theta_0} = \kappa_{x\theta} = \kappa_\theta = 0.$$

If midplane symmetry were imposed:

$$B_{11} = B_{12} = 0$$

If transverse shear deformation were neglected:

$$\epsilon_{xz} = \epsilon_{\theta z} = 0.$$

If the laminate were specially orthotropic then:

$$\overline{Q}_{16_k} = \overline{Q}_{26_k} = 0.$$

If there were no moisture and thermal (hygrothermal) effects:

$$\Delta T = \Delta m = 0.$$

If any or all of these effects were ignored, the expression (6.65) is simplified greatly.

In equation (6.50) the work done by surface forces N_x, N_θ, $N_{x\theta}$ is:

$$\int_{S_T} T_i u_i \, ds = \frac{1}{2} \int_0^{2\pi R} \int_0^L \left\{ N_x \left[\frac{\partial u_0}{\partial x} + \frac{1}{2} \left(\frac{\partial w}{\partial x} \right)^2 \right] \right.$$

$$+ N_\theta \left[\frac{\partial v_0}{\partial s} + \frac{w}{R} + \frac{1}{2} \left(\frac{\partial w}{\partial s} \right)^2 \right]$$

$$\left. + N_{x\theta} \left[\frac{\partial u_0}{\partial s} + \frac{\partial v_0}{\partial x} + 2 \left(\frac{\partial w}{\partial x} \right) \left(\frac{\partial w}{\partial s} \right) \right] \right\} \mathrm{d}x \, \mathrm{d}s$$

where

$$\mathrm{d}s = R \, \mathrm{d}\theta \tag{6.66}$$

If the shell is not a complete cylinder cylinder than its limit $2\pi R$ is replaced by some circumferential dimension.

The form of the lateral displacement, w, in the axial direction for a shell of length L are as follows. In the circumferential direction, similar expressions can be used replacing x by an arc length s and L by a circumferential "length" S_0.

Simple – Simple

$$w = A \sin \frac{m\pi x}{L} \tag{6.67}$$

Simple – Free

$$w = Ax \tag{6.68}$$

Clamped – Clamped

$$w = A \left[1 - \cos \left(\frac{2m\pi x}{L} \right) \right] \tag{6.69}$$

Clamped – Free

$$w = Ax^2 \tag{6.70}$$

Clamped − Simple

$$w = A[L^3x - 3Lx^3 + 2x^4] \tag{6.71}$$

Free − Free

$$w = A \tag{6.72}$$

Then compatible functions must be found for u_0, v_0, γ_x, γ_θ of Equation (6.56) to match the boundary conditions of the problems.

The Minimum Potential Energy expression of this Section was formulated for Lockheed Palo Alto Research Laboratory, and appreciation for their support is hereby expressed.

6.7. Viscoelastic Effects

Recently, Wilson and Vinson [14] have utilized the Theorem of Minimum Potential Energy to ascertain viscoelastic effects on the buckling of composite material plates.

6.8. Problems

6.1. Consider a composite beam of flexural stiffness bD_{11}, *clamped* on each end and subjected to a *uniform* lateral load of $q(x) = q_0$ lbs/in. Use minimum potential energy (MPE) and an assumed deflection function of

$$w(x) = A\left[1 - \cos\frac{2\pi x}{L}\right]$$

to find:
(a) the magnitude and location of the maximum deflection
(b) the magnitude and location of the maximum stress
(c) Does the form of the deflection function satisfy all of the boundary conditions necessary to use MPE for this problem?

6.9 References

1. Vinson, J.R. "Structural Mechanics: The Behavior of Plates and Shells", J. Wiley and Sons, New York, 1974.
2. Vinson, J.R. and T.W. Chou, "Composite Materials and Their Use in Structures" Applied Science Publishers, London, 1975.
3. Sloan, J.G., "The Behavior of Rectangular Composite Material Plates Under Lateral and Hygrothermal Loads", MMAE Thesis, University of Delaware, 1979.
4. Pipes, R.B., J.R. Vinson and T.W. Chou, "On the Hygrothermal Response of Laminated Composite Systems", Journal of Composite Materials, April 1976, pp. 130–148.
5. Wu, C.I. and J.R. Vinson, "Nonlinear Oscillations of Laminated Specially Orthotropic Plates with Clamped and Simply Supported Edges", Journal of the Acoustical Society of America, 49, Part 2, May 1971, pp. 1561–1567.
6. Flaggs, D.L., "Elastic Stability of Generally Laminated Composite Plates Including Hygrothermal Effects", MMAE Thesis, University of Delaware, 1978.
7. Smith, A.P., "The Effect of Transverse Shear Deformation on the Elastic Stability of Orthotropic Plates Due to In-plane Loads", MMAE Thesis, University of Delaware, 1973.
8. Linsenmann, D.R., "Stability of Plates of Composite Materials", MMAE Thesis, University of Delaware, 1974.
9. Warburton, G., "The Vibration of Rectangular Plates", Proceedings of the Institute of Mechanical Engineers, 168, 371 (1954).
10. Young, D. and R. Felgar, Jr., "Tables of Characteristic Functions Representing Normal Modes of Vibration of a Beam", University of Texas Publication No. 4913, July 1944.
11. Felgar, R., Jr., "Formulas for Integrals Containing Characteristic Functions of a Vibrating Beam", Bureau of Engineering Research, University of Texas, 1950.
12. Sloan, J.G. and J.R. Vinson, "The Behavior of Rectangular Composite Material Plates Under Lateral and Hygrothermal Loads", ASME Paper 78 WA/Aero 5.
13. Shen, C. and G.S. Springer, "Moisture Absorption and Desorption of Composite Materials", Journal of Composite Materials, Vol. 10, January 1976, p. 2–20.
14. Wilson, D.W. and J.R. Vinson, "Viscoelastic Analysis of Laminated Plate Buckling Including Transverse Shear and Normal Deformations, AIAA Journal, Vol. 22, July 1984, pp. 982–988.

7. Strength and Failure Theories

7.1. Introduction

In the past sections we have focused upon the functional requirements of beam, plate and shell structural elements as subjected to particular loading environments. The material presented in this section now addresses the broader based objective of having acquired a knowledge of the load analysis methodology, how can one now apply this knowledge to design a structural system to ensure a potentially safe design.

This objective, for composite structures design, is captured in the closed loop stress design profile shown below in Figure 7.1. It should be noted that as with any structural design, a key feature in the design is the selection of an appropriate failure criterion and the inherent iterative nature of the design process proper. In addition, for the case of monolithic materials such as metals it is sufficient to use one observable metric such as the ultimate tensile, compressive, or shear stress to describe failure. For composites, however, the structural engineer is faced with the dilemma of selecting a suitable failure criterion based upon a number of observable stress metrics.

Thus one of the most difficult and challenging subjects to which one is exposed to in the mechanics of advanced composite materials involves finding a suitable failure criterion for these systems. This is compounded by our still limited grasp of understanding and predicting adequately the failure of all classes of monolithic materials to which we seek recourse for

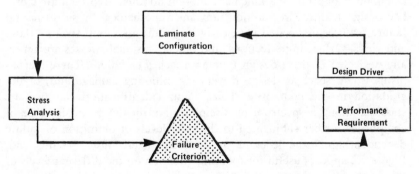

Figure 7.1.

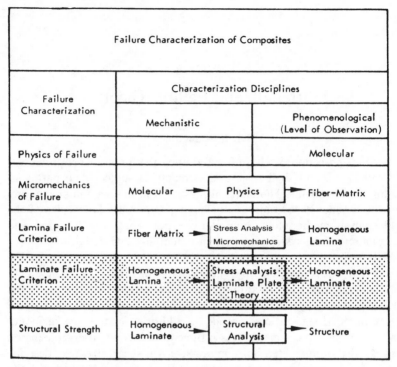

Figure 7.2. Classification of composite failure characterization disciplines.

guidance in establishing descriptive failure criterion for composites. To a large degree the development of such criteria must be associated with philosophical notions of what the concept of failure is about. For example, in most instances failure is perceived to be separation of structural components or material parts of components. This of course need not be the case since the function of the material and/or component may be the design driver and thus, for example, excessive wear in an axle joint may produce a kinematic motion no longer representative of a key design feature. In addition, any micromechanical or substructural failure features associated with early on failure initiation such as flaws and voids, surface imperfections, and built-in residual stresses are generally neglected in these design type approaches to failure. These mechanisms would serve as design drivers for initiating damage and/or degradation in the composite. Thus, failure identification can best be classified along a spectrum of disciplines and in the particular case of composites further subjugated to different levels of definition of failure dependent upon the level of material characterization. To this end, Figure 7.2 appears useful for focusing attention on the different levels of failure characterization and discipline linkage necessary for identity with each failure type. We focus attention in this section on laminate failure criteria which are based upon lamina and micromechanical failure initiators. We do not address in this section the more global problem of failure at joints and attachments.

At this point it is worthwhile to identify the important features and mechanisms controlling the microfailure of composite systems which produce the damage and/or degradation leading to failure in composite laminates. First, it should be emphasized that the matrix and reinforcing fibers, the primary constituents of a composite, have in general widely different strength characteristics. Also, the interface between the fiber and matrix is known to exhibit a response behavior different than that of the bulk matrix in general. The presence of flaws or defects, introduced into the material system during fabrication, may act as stress raisers or failure initiators. Thus, any approach to constructing a micromechanical strength theory should by necessity take into account the influence of such factors. Keeping these thoughts in mind, it appears feasible to characterize failure at the micro level by introducing such local failure modes as

| Fiber dominated failure |

(Breakage, microbuckling, dewetting)

| Bulk matrix dominated failures |

(Voids, crazing)

| Interface/flaw dominated failures |

(Crack propagation, edge delamination)

Once again, such factors while being important failure initiators, are not considered as the primary or causal factors when discussing failure in a global sense. Therefore, as engineers we seek out an observational level of failure to which we can readily relate and feel comfortable with in describing the appropriate failure mechanisms. In this regard we generally attempt to specify lamina failure for anisotropic unidirectional composites or alternatively lamina as existing in composite orthotropic laminates.

To begin this discussion, it has been observed that failure of a laminate consisting of a number of plies oriented in different directions relative to the loading configuration occurs gradually. This occurs due to the fact that an individual lamina failure causes a redistribution of stresses within the remaining laminae of the laminate. Although the failure modes can be either fiber dominated, matrix dominated or interface/flaw dominated, these basic mechanisms would appear as more important features to material scientists or material designers. For structural designers, however, the important element to consider is the lamina, made of a chosen fiber-matrix system, which is necessary for strength and failure analysis.

Before describing the failure of isotropic materials, as a precursor to establishing failure theories for composites, it is worthwhile to indicate some relevant and distinctive features concerning the strength characteristics of composite materials. A composite lamina is known to exhibit an anisotropic strength behavior, that is, the strength is directionally dependent and thus its tensile and compressive strengths are found to be widely different. In addition, the orientation of the shear stresses with respect to the fiber direction in the lamina has been observed to have a significant influence on its strength. Finally, another consideration that merits attention for laminate failure, is that the final fracture depends not only on a failure mode or the number of interactive failure modes occurring but also depends on the failure mode which dominates the failure process. These key features of the strength and failure of composites make the subject matter both complex and challenging.

Thus, while we are able to describe the failure of isotropic materials by an allowable stress field associated with the ultimate tensile, compressive and/or shear strength of the material, the corresponding anisotropic (orthotropic) material requires knowledge of at least five principle stresses. These are the longitudinal tensile and compressive stresses, the transverse tensile and compressive stresses, and the corresponding shear stress. In order to attack a failure problem in composites, however, recourse to monolithic material or metals failure criterion are relied upon to serve as role models for both modification and suitability in predicting failure of laminates. It is therefore worthwhile to review some of the classical failure criteria associated with homogeneous isotropic materials to serve as a baseline reference to development of appropriate criteria for the case of failure of homogeneous anisotropic or orthotropic single-ply lamina. Generally speaking, these combined loading failure (strength) theories can be categorized into three basic types of criteria as follows:

- Stress Dominated
- Strain Dominated
- Interactive

Geometrically speaking, the above criteria can be interpreted in terms of so-called failure envelopes as noted in the accompanying Table 7.1.

Further, each of the failure criteria referred to in Table 7.1 can be redefined in either stress or strain space according to specific tabular schemes. We first examine, however, some general comments as related

Table 7.1. Geometric Interpretation of Failure

• One Stress Metric – Point Failure Envelope
• Two Stress Metric – Two Dimensional Failure Envelope
• N dimensional Stress Metric – N Dimensional Failure Envelope

to classical developments of established failure criteria for metals and then proceed to examine these criteria further in the case of composites.

7.2. Failure of Monolithic Isotropic Materials

The ensuing brief discussion in which failure as defined by the occurence of yielding or fracture is introduced in an attempt to establish a linkage to failure criteria used for anisotropic materials. This enables the reader to focus attention on the use of inductive approaches, based upon well founded contributions, for establishing innovative methodologies in new disciplines. To this end, one of the earliest failure criteria * was suggested by Rankine (1858) who proposed a theory for the yielding of homogeneous isotropic materials having unequal tensile and compressive yield strength values.

This theory known as the Maximum Principal Stress Theory simply states that when any state of stress exists in a structural element which exceeds the yield strength of the material in simple tension or compression then failure occurs. Stated mathematically in terms of principal stresses for the case of unequal tensile and compressive yield strengths in the general three dimensional case can be given as:

$$\sigma_{11} \leqslant \sigma_{yp}^{T} \quad \sigma_{11} \leqslant \sigma_{yp}^{C}$$
$$\sigma_{22} \leqslant \sigma_{yp}^{T} \quad \sigma_{22} \leqslant \sigma_{yp}^{C} \qquad (7.1)$$
$$\sigma_{33} \leqslant \sigma_{yp}^{T} \quad \sigma_{33} \leqslant \sigma_{yp}^{C}$$

For two-dimensional stresses, $\sigma_{33} = 0$ then,

$$\sigma_{11} \leqslant \sigma_{yp}^{T} \quad \sigma_{11} \leqslant \sigma_{yp}^{C}$$
$$\sigma_{22} \leqslant \sigma_{yp}^{T} \quad \sigma_{22} \leqslant \sigma_{yp}^{C} \qquad (7.2)$$

* Failure is usually defined as being related to material yield or fracture.

If the material has the same yield point in tension and compression then,

$$\sigma_{11} = \pm \sigma_{yp}$$
$$\sigma_{22} = \pm \sigma_{yp} \tag{7.3}$$

The maximum strain criterion states that failure occurs when at any point in a structural element the maximum strain at that point reaches the yield value equal to that occurring in a simple uniaxial tension or compression test. Thus result can be stated mathematically by equating the principal strains using Hooke's Law to the uniaxial strain at yield. This results in the following set of equations in terms of principal stresses:

$$\sigma_{11} - \nu\sigma_{22} - \nu\sigma_{33} = \sigma_{yp}^{T}, \sigma_{yp}^{C}$$
$$\sigma_{22} - \nu\sigma_{33} - \nu\sigma_{11} = \sigma_{yp}^{T}, \sigma_{yp}^{C} \tag{7.4}$$
$$\sigma_{33} - \nu\sigma_{11} - \nu\sigma_{22} = \sigma_{yp}^{T}, \sigma_{yp}^{C}$$

For the case of plane stress, $\sigma_{33} = 0$, the above equations simplify to

$$\sigma_{11} - \nu\sigma_{22} = \sigma_{yp}^{T}, \sigma_{yp}^{C}$$
$$\sigma_{22} - \nu\sigma_{11} = \sigma_{yp}^{T}, \sigma_{yp}^{C} \tag{7.5}$$

Two other important failure criteria for homogeneous isotropic materials are the Maximum Shear Stress Theory often referred to as Tresca's Theory and the Distortion Energy Criteria often referred to as Von Mises Criterion.

The first theory, that is the so-called Maximum Shear Stress Theory, suggests that yielding will occur in a material when the maximum shear stress reaches a critical value, the maximum shear stress at yielding in a uniaxial stress state. Expressed quantitatively in terms of principal stresses this result can be stated as:

$$\sigma_{11} - \sigma_{22} = \pm \sigma_{yp}$$
$$\sigma_{22} - \sigma_{33} = \pm \sigma_{yp} \tag{7.6}$$
$$\sigma_{33} - \sigma_{11} = \pm \sigma_{yp}$$

For the special case of plane stress, when $\sigma_{33} = 0$, the expression simplifies to read:

$$\sigma_{11} - \sigma_{22} = \pm \sigma_{yp}$$
$$\sigma_{22} = \pm \sigma_{yp} \tag{7.7}$$
$$\sigma_{11} = \pm \sigma_{yp}$$

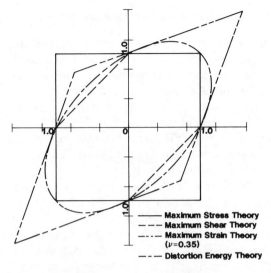

COMPARISON OF FAILURE THEORIES

Figure 7.3. Comparison of failure theories.

Corresponding to Tresca's yield criterion a second popular yield criterion for homogeneous isotropic materials was attributed to Von Mises (1913). This theory known as the distortion energy criterion also assumes that the principal influence on material yield is the maximum shearing stress. In this theory, the elastic energy of a material can be considered to consist of a dilatational and distortional component. It is recognized that the dilatational part of the total work is dependent upon a hydrostatic stress state which does not produce yielding at normal pressures in homogeneous isotropic materials. Thus, the remaining or distortional energy can be expressed in terms of principal stresses and written as:

$$\sigma_{11}^2 + \sigma_{22}^2 + \sigma_{33}^2 - \sigma_{11}\sigma_{22} - \sigma_{22}\sigma_{33} - \sigma_{33}\sigma_{11} = \sigma_{yp}^2 \tag{7.8}$$

A careful examination of this criterion with the previous Maximum Shearing Stress Theory indicates that at most the two theories deviate by no more than fifteen percent.

These four failure criterion depicted in Figure 7.3, while not exhaustive of all of the types proposed for metals, are representative models of the types often used and as such have led to the anisotropic failure criterion highlighted in the following section.

7.3. Anisotropic Strength and Failure Theories

We now focus attention on the development of useful macroscopic yield criteria as extended from metals behavior to the case of composite (anisotropic) materials. A more comprehensive review of this subject and the individual theories can be found in Ref. 6. In the following paragraphs, those theories which have been most often cited in the composites literature are reviewed in terms of their origin and salient features for the reader. This review serves as an important interface between the classical theories previously described and as a means for comparison for various composite systems analytically and experimentally.

As mentioned earlier in the text, a fundamental requirement in the design of structural systems from structural elements such as beams, plates, and shells is a knowledge of the strength and/or failure of such elements subjected to complex loading states. The strength and failure of these structural elements can be associated with the yield strength of the material or correspondingly its ultimate strength. For brittle materials the ultimate or failure strength appears to be a more suitable base line reference value to determine failure while for ductile type materials the yield strength appears a more useful criterion. In the case of anisotropic materials and in particular high performance composites, the fiber reinforcements are generally elastic up to failure while the matrices can be either of the brittle or ductile type. The latter are mainly associated with metal matrix composites (MMC's), while the majority of advanced composite systems use thermosetting (epoxy) matrices for their binders. Thus, all subsequent reference to failure in this chapter as related to high performance composites will be based upon reference to the ultimate strength of the material. This should, of course, be altered if ductile matrices or fiber types other than brittle elastic are used. One other unique feature of anisotropic material failure is that modes of failure are important considerations in the design process. This is due to the fact that uniqueness in yield or failure is associated with the direction of testing.

We now proceed to examine anisotropic failure criteria in the light of the aforementioned remarks, and as extensions of failure theories based upon the three basic types established for monolithic materials, that is,

- Stress Dominated
- Strain Dominated
- Interactive

The historical development, as presented, reflects a somewhat selective one and the reader is referred to Ref. 6 for a more comprehensive discussion. In addition, the reader is cautioned that the theories, as constructed and referred to, do not include environmental factors such as temperature and humidity, and also do not include initial (residual) stresses that are usually present in composites.

7.3.1. Maximum Stress Theory

In doing research in the forest products area, Jenkins (1920) extended the Maximum Stress Theory to orthotropic materials. He stated that failure occurred when one or all of the orthotropic stress values exceed their maximum limits as obtained in uniaxial tension, compression, or pure shear stress tests, when the material is tested to failure. This can be stated analytically as:

$$\sigma_{11} = X$$
$$\sigma_{22} = Y \qquad\qquad (7.9)$$
$$\sigma_{12} = S$$

Further extensions of this theory were presented by Stowell and Liu (1961) [8] and Kelly and Davies (1965) [9].

7.3.2. Maximum Strain Theory

The Maximum Strain Theory states that failure occurs when the strain obtained along the principal material axes exceed their limiting values. Waddoups (1966) [10] utilized these conditions in assessing the failure strength of orthotropic materials expressing the results analytically as

$$\bar{\epsilon}_{11} = \frac{\sigma_{11}}{E_{11}} - \nu_{12}\frac{\sigma_{22}}{E_{11}}$$
$$\bar{\epsilon}_{22} = \frac{\sigma_{22}}{E_{22}} - \frac{\nu_{12}}{E_{11}} \qquad \bar{\gamma}_{12} = \frac{\sigma_{12}}{G_{12}} \qquad (7.10)$$

The three principal strain values can be referred to modes of failure in the corresponding strain directions.

7.3.3. Interactive Theories

One of the earliest interactive failure criterion for anisotropic materials was initiated by Hill (1948). This theory is a generalization of the isotropic yield behavior of ductile metals for the case of large strains. That is, as a metal is strained in a certain direction, for example in a rolling process, the material grains tend to become aligned and a self induced anisotropy occurs. Hill thus formulated an interactive yield criterion for such materials which can be written in terms of the stress components as:

$$F(\sigma_{22} - \sigma_{33})^2 + G(\sigma_{33} - \sigma_{11})^2 + H(\sigma_{11} - \sigma_{22})^2$$

$$+ 2L\sigma_{23}^2 + 2M\sigma_{13}^2 + 2N\sigma_{12}^2 = 1 \tag{7.11}$$

where the quantities F, G, H, L, M, N reflect the current state of material anisotropy.

For unidirectionally reinforced composites $M = N$, $G = H$, we obtain

$$F(\sigma_{22} - \sigma_{33})^2 + G(\sigma_{33} - \sigma_{11})^2 + G(\sigma_{11} - \sigma_{22})^2$$

$$+ 2L\sigma_{23}^2 + 2M(\sigma_{31}^2 + \sigma_{12}^2) = 1. \tag{7.12}$$

For a composite lamina or laminates in a plane state of stress $\sigma_{33} = 0$, the above equation reduces to

$$F\sigma_{22}^2 + G\sigma_{11}^2 + H(\sigma_{11} - \sigma_{22})^2 + 2N\sigma_{12}^2 = 1. \tag{7.13}$$

An extension of Hill's criterion to account for unequal tension and compression for anisotropic materials was introduced by Marin (1956) [11]. Stated in terms of principal stresses, failure (yielding) is assumed to occur when the following condition is satisfied

$$(\sigma_{11} - a)^2 + (\sigma_{22} - b)^2 + (\sigma_{33} - c)^2 + q[(\sigma_{11} - a)(\sigma_{22} - b)$$

$$+ (\sigma_{22} - b)(\sigma_{33} - c) + (\sigma_{33} - c)(\sigma_{11} - a)] = \sigma^2 \tag{7.14}$$

The quantities a, b, c, q, and r are experimentally determined parameters.

Marin's anisotropic failure criterion was further extended by Norris (1962) [12] who introduced nine stress components to define failure. These nine properties consist of three tensile, three compressive and three shear strength values. The equations which must be satisfied for failure to occur are thus given by:

$$\left(\frac{\sigma_{11}}{X}\right)^2 + \left(\frac{\sigma_{22}}{Y}\right)^2 - \frac{\sigma_{11}\sigma_{22}}{XY} + \left(\frac{\sigma_{12}}{S}\right)^2 = 1$$

$$\left(\frac{\sigma_{22}}{Y}\right)^2 + \left(\frac{\sigma_{33}}{Z}\right)^2 - \frac{\sigma_{22}\sigma_{33}}{YZ} + \left(\frac{\sigma_{23}}{T}\right)^2 = 1 \tag{7.15}$$

$$\left(\frac{\sigma_{33}}{Z}\right)^2 + \left(\frac{\sigma_{11}}{X}\right)^2 - \frac{\sigma_{33}\sigma_{11}}{ZX} + \left(\frac{\sigma_{13}}{R}\right)^2 = 1$$

where the quantities X, Y, Z are the tensile or compressive yield strengths of the material while R, S, T are the corresponding shear strengths.

For the case of plane stress the above equations reduce to

$$\left(\frac{\sigma_{11}}{X}\right)^2 + \left(\frac{\sigma_{22}}{Y}\right)^2 - \frac{\sigma_{11}\sigma_{22}}{XY} + \left(\frac{\sigma_{12}}{S}\right)^2 = 1$$

$$\left(\frac{\sigma_{22}}{Y}\right)^2 = 1 \quad \left(\frac{\sigma_{11}}{X}\right)^2 = 1 \tag{7.16}$$

The plane stress results by Hill were simplified for the case of fiber reinforced composites by Azzi and Tsai (1965) [13] considering the composite to be transversely isotropic. Thus with $Z = Y$, the modified form of Hill's plane stress criterion can be written as,

$$\left(\frac{\sigma_{11}}{X}\right)^2 + \left(\frac{\sigma_{22}}{Y}\right)^2 - \left(\frac{\sigma_{11}\sigma_{22}}{X}\right)^2 + \left(\frac{\sigma_{12}}{S}\right)^2 = 1. \tag{7.17}$$

The above result is equivalent to that obtained by Norris with the distinction that the latter's data is somewhat more general in accounting for difference in yield in both tension and compression.

A generalization of this failure criterion to incorporate the effects of brittle materials was considered by Hoffman (1967) [14]. In terms of stress components, failure occurs when the following equations are satisfied:

$$C_1(\sigma_{22} - \sigma_{33})^2 + C_2(\sigma_{33} - \sigma_{11})^2 + C_3(\sigma_{11} - \sigma_{22})^2 + C_4\sigma_{11} - C_5\sigma_{22}$$

$$+ C_6\sigma_{33} + C_7\tau_{23}^2 + C_8\tau_{13}^2 + C_9\tau_{12}^2 = 1. \tag{7.18}$$

Quantities C_1, C_2, C_3, C_4, C_5, C_6, C_7, C_8, C_9 are constants determined from materials properties tests.

$$C_1 = \frac{1}{2}\left[\frac{1}{Y_T Y_C} + \frac{1}{Z_T Z_C} - \frac{1}{X_T X_C}\right]$$

$$C_2 = \frac{1}{2}\left[\frac{1}{Z_T Z_C} + \frac{1}{X_T X_C} - \frac{1}{Y_T Y_C}\right] \tag{7.19}$$

$$C_3 = \frac{1}{2}\left[\frac{1}{X_T X_C} + \frac{1}{Y_T Y_C} - \frac{1}{Z_T Z_C}\right]$$

$$C_4 = \frac{1}{X_T} - \frac{1}{X_C} \quad C_7 = \frac{1}{T^2}$$

$$C_5 = \frac{1}{Y_T} - \frac{1}{Y_C} \quad C_8 = \frac{1}{R^2}$$

$$C_6 = \frac{1}{Z_T} - \frac{1}{Z_C} \quad C_9 = \frac{1}{S^2}$$

This theory also incorporates unequal tensile and compressive failure strengths as an inherent part of the development. For the case of plane stress this equation reduces to:

$$\frac{\sigma_{12}^2}{X_T X_C} + \frac{\sigma_{22}^2}{Y_T Y_C} - \frac{\sigma_{11}\sigma_{22}}{X_T X_C} + \frac{X_C - X_T}{X_T X_C}\sigma_{11} + \frac{Y_C - Y_T}{Y_T Y_C}\sigma_{22} + \frac{\sigma_{12}^2}{S^2} = 1 \quad (7.20)$$

A generalization of the Hoffman result to incorporate a more comprehensive definition for failure was later proposed by Tsai and Wu (1971) [15]. In this criterion failure is assumed to occur when the following equations are satisfied:

$$F_i\sigma_i + F_{ij}\sigma_i\sigma_j = 1 \quad (i, j = 1, 2, \dots, 6)$$
$$F_{ii}F_{jj} - F_{ij}^2 \geqslant 0 \quad (i, j = 1, 2, \dots, 6) \quad (7.21)$$

The quantities F_i and F_j are related to the tensile and compressive yield strengths of the material while six shear tests both positive and negative are required to define the parameters F_1 through F_6 and F_{44} through F_{66}.

$$F_1 = \frac{1}{X_T} - \frac{1}{X_C} \quad F_{11} = \frac{1}{X_T X_C}$$

$$F_2 = \frac{1}{Y_T} - \frac{1}{Y_C} \quad F_{22} = \frac{1}{Y_T Y_C}$$

$$F_3 = \frac{1}{Z_T} - \frac{1}{Z_C} \quad F_{33} = \frac{1}{Z_T Z_C}$$

$$F_4 = \frac{1}{T^+} - \frac{1}{T^-} \quad F_{44} = \frac{1}{T^+ T^-}$$

$$F_5 = \frac{1}{R^+} - \frac{1}{R^-} \quad F_{55} = \frac{1}{R^+ R^-}$$

$$F_6 = \frac{1}{S^+} - \frac{1}{S^-} \quad F_{66} = \frac{1}{S^+ S^-}$$

The ultimate test for any of the criterion as discussed above is based upon its correlation with available test data. At the present time an adequate number of tests has not been made to ascertain any deterministic values for comparative purposes. However, a recent survey of failure criterion which have been proposed for use in practice have been featured in Ref. [16]. The results of this survey indicate that the following failure criteria appear most popular in practical use today:

- Maximum Strain
- Maximum Stress
- Interactive Criteria
  - Tsai-Hill
  - Tsai-Wu
  - Chamis
  - Hoffman

A number of these theories will be used in selected examples in the latter part of this chapter. We begin our discussion of failure criterion by first examining failure of singly-ply unidirectional fiber composites and progress to multiply composites consisting of variable angle ply geometries and orientations with respect to complex stress fields.

7.4. Lamina Strength Theories

We first examine the single ply strength of uniaxial lamina subjected to tensile loads. At the level of analysis being discussed and due to the success of micromechanical principles in predicting the elastic properties of composites, similar principles have been advanced for predicting the overall strength of composites as well. The problem with using this approach is that in applying a deterministic approach to the case where probabilistic events predominate, we generally obtain an invalid result. This is due to the wide variety of fiber and/or fiber bundle dominated strengths exhibited in composite networks. Further, the effects of these load perturbations can result in high stress concentrations in the vicinity of the newly failed fibers or be propagated/dispersed to remote locations in the composite. This results in a failure model which is both dependent on the relative strength levels in the different regions of the composite as well as involvement of interactive failure modes in a broad sense. Thus in the end, a requirement for well-planned experiments appears to be a necessary building block for understanding and predicting such failure phenomena. As an example, in the case of compressive loads both material and configurational failures need to be addressed. That is both the compressive strength and buckling strengths need to be considered. Local fiber instabilities identified by in-phase and out-of-phase buckling (1), and fiber eccentricities (2) also need to be considered. The corresponding transverse tension and compression loadings are generally dominated by the so-called matrix mode of failure. In the case of compressive transverse loads, there appears to be several microscopic shear failure modes active, these being in the plane of and perpendicular to the filaments. Longitudinal and transverse shear stresses in general also result in matrix dominated failure modes although there is some theoretical evidence to suggest an increase in the in-plane shear stress (3).

Since most structural components are subjected to complex loading states the question of what happens to structural elements under combined loads is an important question. In the preceding qualitative discussion of strength investigation for single-ply laminae it was mentioned that material characterization was possible by obtaining data through relatively simple uniaxial and shear tests. For the case of composites, a systematic experimental program to investigate failure for all possible complex stress states becomes prohibitive and recourse to other predictive methodologies appears necessary.

In order to establish guidelines for strength and failure of laminae which are the building blocks of laminates, it is necessary to consider a number of fundamental features related to composite material behavior. To this end, it is important to point out that while we recognize the inherent microlevel mechanisms associated with failure in laminae we only consider failure to occur at the macrolevel. This observation is predicated to a large extent by the current state of the art in failure analysis which precludes an ability to quantitatively relate in any interactive fashion events occurring at the micro and macro level of analysis. Thus, for design purposes the engineer must resort to consideration of strength degradation as observed for laminae and relate these data subsequently to laminates.

Further, since we deal with laminae as our fundamental building units in composites we characterize their properties through means of either analytical methodologies (micromechanics) or experimental (empirical) data. The data as shown in Figures 7.4 and 7.5 serve as a means for obtaining the necessary uniaxial stress and strain allowables in tension, compression, and shear. Since most laminae can be considered as either transversely isotropic or specially orthotropic, we generally need only four independent elastic compliances or stiffnesses and five strengths to define the material system.

Again, as mentioned previously a single lamina is considered to be in a plane state of stress and principal strengths are used as comparative strength criteria.

- Maximum Stress
- Maximum Strain
- Interactive Laws

Each of these theory types will be used in testing for the failure of laminae subjected to several states of stress.

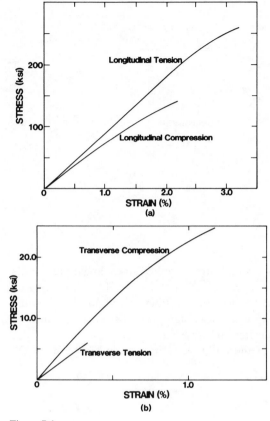

Figure 7.4.

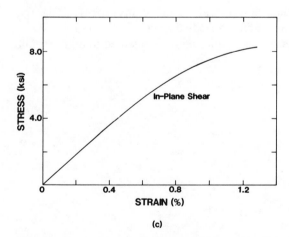

Figure 7.5. Typical stress-strain curves for S-glass fibers in an epoxy matrix.
(a) Unidirectional laminate with $E_{11} = 8.8 \times 10^6$ psi for longitudinal tension and $E_{11} = 7.2 \times 10^6$ psi for longitudinal compression.
(b) Unidirectional laminate with $E_{12} = 2.5 \times 10^6$ psi for transverse compression and $E_{12} = 1.9 \times 10^6$ psi for transverse tension.
(c) In-plane shear with $G_{12} = 0.8 \times 10^6$ psi.

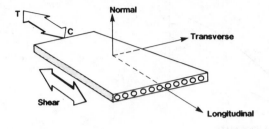

SINGLE LAMINA

Figure 7.6.

Example 1.

We begin by examining a single stress applied to an angle ply lamina material made of *S* glass/epoxy, with the stress, $\sigma_x = 500$ psi, making an angle, $\theta = 60°$, with respect to the principal fiber directions. This loading is shown in Figure 7.7 along with the design properties corresponding to this material system as indicated in Table 7.2.

We now examine each of the failure laws in turn and in terms of the loading shown.

Maximum Stress

This theory states that failure will occur if any stress along the principal directions of the lamina exceeds the specified allowables. Analytically the following set of inequalities must be satisfied to ensure a fail safe design.

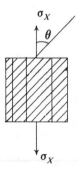

Figure 7.7.

Table 7.2. *S* Glass/Epoxy

$E_L = E_X = 8.42 \times 10^6$ psi
$E_T = E_Y = 2.00 \times 10^6$ psi
$G_{LT} = 0.77 \times 10^6$ psi
$\nu_{LT} = 0.293$
$\nu_{LT} = 0.067$
280,000 psi $= \sigma_L^C$
136,000 psi $= \sigma_L^T$
  4,000 psi $= \sigma_T^T$
 20,000 psi $= \sigma_T^C$
  6,000 psi $= \sigma_{LT}^S$
   0.67 psi $= V_f$

Tension	Compression
$\sigma_L < \sigma_{LU}^t$	$\sigma_L < \sigma_{LU}^c$
$\sigma_T < \sigma_{TU}^t$	$\sigma_T < \sigma_{TU}^c$
$\sigma_{LT} < \sigma_{LTU}$	

For the present example and in order to test the failure theory, it is first necessary to transform the given stress σ_x along the direction of the principal material direction. Using the transformation equations we can thus write in a general form the following results:

$$\sigma_L = \sigma_x \cos^2 \theta = 125 \text{ psi}$$

$$\sigma_T = \sigma_x \sin^2 \theta = 375 \text{ psi}$$

$$\sigma_{LT} = \sigma_x \sin \theta \cos \theta = 216.5 \text{ psi}$$

Substituting the given values for σ_x and θ we can test the computed stresses in the material directions against the allowable stresses obtaining,

$$\sigma_L = 125 < \sigma_{LU} = 280,000$$

$$\sigma_T = 375 < \sigma_{TU} = 4,000$$

$$\sigma_{LT} = 216.5 < \sigma_{LTU} = 6,000$$

The results obtained can also be cast in terms of non-dimensional stress values and plotted as shown in Figure 7.8.

Maximum Strain

The maximum strain theory states that failure occurs when the strain along the material axes exceeds the allowable strain in the same principal

226

direction as obtained in the same mode of testing. That is,

Tension		Compression
$\epsilon_L < \epsilon_{LU}^t$		$\epsilon_L < \epsilon_{LU}^c$
$\epsilon_T < \epsilon_{TU}^t$		$\epsilon_T < \epsilon_{TU}^c$
	$\gamma_{LT} < \gamma_{LTU}$	

Once again, for the orthotropic lamina subjected to a load σ_X, we make use of the transformed stresses as found previously in the principal material coordinate directions. In order to calculate the corresponding strains in the principal directions (material coordinates) we use Hooke's Law for orthotropic materials. Therefore, we can write:

$$\epsilon_L = \frac{1}{E_L} \left[\cos^2 \theta - \nu_{LT} \sin^2 \theta \right] \sigma_X$$

$$\epsilon_T = \frac{1}{E_T} \left[\sin^2 \theta - \nu_{TL} \cos^2 \theta \right] \sigma_X$$

$$\gamma_{LT} = \frac{1}{G_{LT}} \left[\sin \theta \cos \theta \right] \sigma_X$$

Substituting the given values of σ_X, θ, ν_{LT}, E_L, E_T and G_{LT} from Table 7.2 we determine ϵ_L, ϵ_T and γ_{LT}. These values are computed and tested

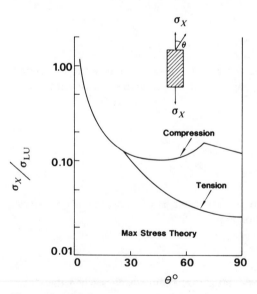

Figure 7.8. Analysis of an orthotropic lamina.

against the allowable strain values which are given below,

$\epsilon_L < \epsilon_{LU}$ $\epsilon_L = 0.000001 < \epsilon_{LU} = 0.033$

$\epsilon_T < \epsilon_{TU}$ $\epsilon_T = 0.000183 < \epsilon_{TU} = 0.002$

$\gamma_{LT} < \gamma_{LTU}$ $\gamma_{LT} = 0.000281 < \gamma_{LTU} = 0.0078$

The Maximum Strain Theory can also be graphically depicted and compared with the Maximum Stress Theory for the example cited. The differences in the theoretical predictions are small, this due to the fact that the material is considered to behave as linearly elastic to failure and consequently is not shown.

Interactive Theory

The third theory is of the interactive type, and states that failure occurs under some combined (multiplicative and/or additive) set of stresses. For the current examples, we select one of the interactive criteria for use specifically the Azzi/Tsai type. This failure criterion can be stated analytically for the case of plane stress as:

$$\left(\frac{\sigma_L}{\sigma_{LU}}\right)^2 - \left(\frac{\sigma_L}{\sigma_{LU}}\right)\left(\frac{\sigma_T}{\sigma_{LU}}\right) + \left(\frac{\sigma_T}{\sigma_{TU}}\right)^2 + \left(\frac{\sigma_{LT}}{\sigma_{LTU}}\right)^2 < 1$$

In the present example for an applied stress σ_X, the above equation can be written as:

$$\left(\frac{\cos^2\theta}{\sigma_{LU}}\right)^2 - \left(\frac{\cos\theta\sin\theta}{\sigma_{LU}}\right)^2 + \left(\frac{\sin^2\theta}{\sigma_{TU}}\right)^2 + \left(\frac{\sin\theta\cos\theta}{\sigma_{LTU}}\right)^2 < \frac{1}{\sigma_X^2}$$

Using the numbers given in Table 1 this equation is written as:

$$\left(\frac{0.25}{280,000}\right)^2 - \left(\frac{0.433}{280,000}\right)^2 + \left(\frac{0.75}{4,000}\right)^2 + \left(\frac{0.433}{6,000}\right)^2 < \frac{1}{(500)^2}$$

As was done in the case of the Maximum Stress Theory, a similar graphical plot of the off axis composite strength can be presented. This is shown in Figure 7.9 for comparison with the Maximum Stress (Strain) Theory. A comparison of these theories leads to the fact that the Maximum Stress Theory as does the Maximum Strain predicts somewhat higher strength values. The major discrepancy between the theories occurs in the angle at which a transition from one failure mode to an alternative mode occurs (compare curves).

228

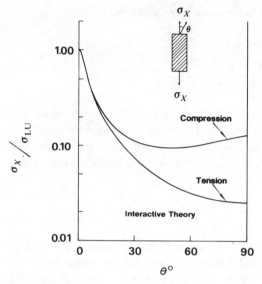

Figure 7.9.

Example 2.

As a second example we consider the lamina composite shown in Figure 7.10 considering the material properties as given in Table 2.

Maximum Stress Theory

As stated previously, this theory predicts failure when anyone of the ultimate (yield) strength values exceeds the corresponding allowable stresses in a principal axes direction. Thus, in order to ensure a safe design once again the following inequalities must be satisfied in both tension and compression.

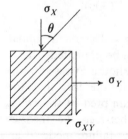

Figure 7.10.

Tension	Compression
$\sigma_L \leqslant \sigma'_{LU}$	$\sigma_L \leqslant \sigma^c_{LU}$
$\sigma_T \leqslant \sigma'_{TU}$	$\sigma_T \leqslant \sigma^c_{TU}$
$\sigma_{LT} \leqslant \sigma_{LTU}$	

In the example given, the stresses as shown must be translated from the geometric axes to the principal material directions. This can be done through equilibrium considerations on the element or use of the transformation equations presented previously. Thus we can use

$$\sigma_{11} = \sigma_X \cos^2 \theta + \sigma_Y \sin^2 \theta + 2\sigma_{XY} \cos \theta \sin \theta$$

$$\sigma_{22} = \sigma_X \sin^2 \theta + \sigma_Y \cos^2 \theta - 2\sigma_{XY} \cos \theta \sin \theta$$

$$\sigma_{12} = -\sigma_X \sin \theta \cos \theta + \sigma_Y \sin \theta \cos \theta + \sigma_{XY}(\cos^2 \theta - \sin^2 \theta)$$

or the transformation equations

$$\begin{bmatrix} \sigma_{11} \\ \sigma_{22} \\ \sigma_{12} \end{bmatrix} = \begin{bmatrix} T \end{bmatrix} \begin{bmatrix} \sigma_X \\ \sigma_Y \\ \sigma_{XY} \end{bmatrix}$$

Using the latter and substituting $\theta = 60°$ into the transformation matrix we obtain

$$\begin{bmatrix} \sigma_{11} \\ \sigma_{22} \\ \sigma_{12} \end{bmatrix} = \begin{bmatrix} 1/4 & 3/4 & \sqrt{3}/2 \\ 3/4 & 1/4 & -\sqrt{3}/2 \\ -\sqrt{3/4} & \sqrt{3/4} & -1/2 \end{bmatrix} \begin{bmatrix} -500 \\ 1000 \\ -200 \end{bmatrix} = \begin{bmatrix} 451.8 \\ 48.2 \\ 749.5 \end{bmatrix}$$

Comparing the stresses calculated in the right-hand column vector with the allowable values indicates that the lamina is fail safe, that is

$$\sigma_L < \sigma'_{LU}$$

$$\sigma_T < \sigma'_{TU}$$

$$\sigma_{LT} < \sigma'_{LTU}$$

Importance of Shear Stresses

Some comment on the role of the shear stress in determining the strength of lamina needs to be addressed. For homogeneous isotropic materials the direction of the shear stress can be either positive or negative and is of little consequence in determining the strength of such materials. These simplistic arguments do not carry over to orthotropic lamina and composites. Consider the lamina consisting of $(-45°)$ ori-

230

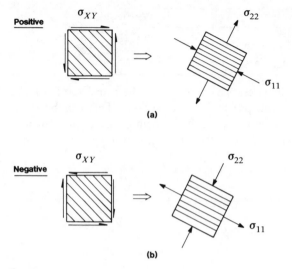

(a)

(b)

Figure 7.11

ented fibers as shown in Fig. 7.11, and loaded by means of positive and negative shear stresses respectively, the signs determined from generally accepted strength of materials conventions. It can be seen that each of these lamina leads to important differences in loads along the principal or material coordinate axes. In particular, for the case of positive shear tensile stresses are developed in the transverse direction and compressive stresses in the fiber direction. For the case of negative shear the reverse loading situation occurs.

Since shear strength of lamina are controlled by the transverse strength of the composite, we can see that for the case of positive shear and negative fiber orientation we are led to lower predictions of apparent strength than when the reverse case prevails. Thus, the off-axis shear of lamina must be carefully examined in the light of the applied direction of the shear stress.

Maximum Strain Theory

This theory states that a fail safe design will exist when the following inequalities are satisfied:

Tension	Compression
$\epsilon_{11} \leqslant \epsilon_{11}^{T}$	$\epsilon_{11} \leqslant \epsilon_{11}^{C}$
$\epsilon_{22} \leqslant \epsilon_{22}^{T}$	$\epsilon_{22} \leqslant \epsilon_{22}^{C}$
$\epsilon_{12} \leqslant \epsilon_{12}$	

The Maximum Strain Theory is directly analogous to the Maximum Stress Theory in terms of correspondence in that stresses are simply replaced by allowable strain values in the fail safe procedure. This is predicated upon the assumption that the material remains elastic up to failure and thus one obtains the allowable strain from:

$$\epsilon_{LU} = \frac{\sigma_{LU}}{E_L} \qquad\qquad \epsilon'_{LU} = \frac{\sigma_{LU}}{E_L}$$

$$\epsilon_{TU} = \frac{\sigma_{TU}}{E_T} \quad \gamma_{LTU} = \frac{\sigma_{LTU}}{G_{LT}} \quad \epsilon'_{TU} = \frac{\sigma'_{TU}}{E'_T}$$

The strains in the L, T directions can be calculated from the Hooke's law relations using the calculated stress values for σ_L, σ_T, and σ_{LT} as input values, that is:

$$\epsilon_L = \frac{\sigma_L}{E_L} - \nu_{TL}\frac{\sigma_T}{E_T} = 0.000052$$

$$\epsilon_T = \frac{\sigma_T}{E_T} - \nu_{LT}\frac{\sigma_L}{E_L} = 0.0000084$$

$$\gamma_{LT} = \frac{\sigma_{LT}}{G_{LT}} = 0.00097$$

Comparing the above values with the allowable strains we see that:

$$\epsilon_L < \epsilon_{LU}$$

$$\epsilon_T < \epsilon_{TU}$$

and,

$$\gamma_{LT} < \gamma_{LTU}$$

Interactive Theory

Two key features of the Maximum Stress and Maximum Strain failure theories should be noted. These are:
- Interaction between strengths is not accounted for
- Failure is dictated by a governing inequality

The first of the above features is sometimes considered an inadequacy of these types of failure theorems and thus one attempts to define involved composite failure theories based upon homogeneous isotropic metal based systems. Due to the large number of such theories available, as discussed previously, only one of these theories will be presented and used in the present problem. The specific failure theory selected is based upon an extension of Hill's generalization of the von Mises or Distortional Energy failure theory which was further extended by Azzi and Tsai (1965). In functional form this equation is of the type,

$$\left(\frac{\sigma_L}{\sigma_{LU}}\right)^2 - \left(\frac{\sigma_L}{\sigma_{LU}}\right)\left(\frac{\sigma_T}{\sigma_{LU}}\right) + \left(\frac{\sigma_T}{\sigma_{TU}}\right)^2 + \left(\frac{\sigma_{LT}}{\sigma_{LTU}}\right)^2 < 1$$

For the present example using the calculated results for σ_L, σ_T, σ_{LT}, and the data included in Table 7.2, we can write,

$$\left(\frac{3160}{280,000}\right)^2 + \left(\frac{340}{4000}\right)^2 - \frac{(3160)(340)}{(280,000)^2} + \left(\frac{5240}{6000}\right)^2 < 1$$

The Failure Criteria described in the preceding paragraphs can be summarized in terms of the three types mentioned earlier in the chapter. That is stress dominated, strain dominated, and so-called stress interactive types. A graphical interpretation of these yield criteria for the case of equal tensile and compressive yield points can be graphically displayed, while an analytical description couched in terms of either strain or stress space follows in the accompanying tables.

Table 7.3.

FAILURE CRITERIA
(Theories with and without Independent Failure Modes)
Maximum Stress Criterion

Stress Space	Strain Space
$\sigma_{LU}^1 \leqslant \sigma_1 \leqslant \sigma_{LU}$	$\epsilon_1 = -\dfrac{C_{12}}{C_{11}}\epsilon_2 - \dfrac{C_{16}}{C_{11}}\epsilon_6 + \dfrac{\sigma_{LU}}{C_{11}}$
$\sigma_{TU}^1 \leqslant \sigma_2 \leqslant \sigma_{TU}$	$\epsilon_1 = -\dfrac{C_{12}}{C_{11}}\epsilon_2 - \dfrac{C_{16}}{C_{11}}\epsilon_6 - \dfrac{\sigma'_{LU}}{C_{11}}$
$\sigma_{LTU}^1 \leqslant \sigma_6 \leqslant \sigma_{LTU}$	$\epsilon_3 = -\dfrac{C_{12}}{C_{22}}\epsilon_1 - \dfrac{C_{26}}{C_{22}}\epsilon_6 + \dfrac{\sigma_{TU}}{C_{22}}$
	$\epsilon_4 = -\dfrac{C_{12}}{C_{22}}\epsilon_1 - \dfrac{C_{26}}{C_{22}}\epsilon_6 - \dfrac{\sigma'_{TU}}{C_{22}}$
	$\epsilon_6 = -\dfrac{C_{16}}{C_{66}}\epsilon_1 - \dfrac{C_{26}}{C_{66}}\epsilon_2 + \dfrac{\sigma_{LTU}}{C_{66}}$
	$\epsilon_6 = -\dfrac{C_{16}}{C_{66}}\epsilon_1 - \dfrac{C_{26}}{C_{66}}\epsilon_2 - \dfrac{\sigma_{LTU}}{C_{66}}$

Table 7.4.

MAXIMUM STRAIN CRITERION

Strain Space	Stress Space
$\epsilon_{LU} = S_{11}\sigma_{LU} \geqslant \epsilon_1$	$\sigma_1 = -\dfrac{S_{12}}{S_{11}}\sigma_2 - \dfrac{S_{16}}{S_{11}}\sigma_6 + \sigma_{LU}$
$\epsilon'_{LU} = S_{11}\sigma'_{LU} \geqslant -\epsilon_1$	$\sigma_1 = -\dfrac{S_{12}}{S_{11}}\sigma_2 - \dfrac{S_{16}}{S_{11}}\sigma_6 - \sigma'_{LU}$
$\epsilon_{TU} = S_{22}\sigma_{TU} \geqslant \epsilon_2$	$\sigma_2 = -\dfrac{S_{12}}{S_{22}}\sigma_1 - \dfrac{S_{26}}{S_{22}}\sigma_6 + \sigma_{TU}$
$\epsilon'_{TU} = S_{22}\sigma'_{TU} \geqslant -\epsilon_2$	$\sigma_2 = -\dfrac{S_{12}}{S_{22}}\sigma_1 - \dfrac{S_{26}}{S_{22}}\sigma_6 - \sigma'_{TU}$
$\epsilon_{LTU} = S_{66}\sigma_{LTU} \geqslant \epsilon_6$	$\sigma_6 = -\dfrac{S_{16}}{S_{66}}\sigma_1 - \dfrac{S_{26}}{S_{66}}\sigma_2 + \sigma_{LTU}$
$\epsilon'_{LTU} = S_{66}\sigma'_{LTU} \geqslant -\epsilon_6$	$\sigma_6 = -\dfrac{S_{16}}{S_{66}} - \dfrac{S_{26}}{S_{66}} - \sigma'_{LTU}$

Table 7.5.

INTERACTIVE FAILURE CRITERION

Stress Space	Strain Space
$f(\tau_i) = F_i\sigma_i + F_{ij}\sigma_i\sigma_j$ $+ F_{ijk}\sigma_i\sigma_j\sigma_k + \ldots$	$g(\epsilon_i) = G_i\epsilon_i + G_{ij}\epsilon_i\epsilon_j$ $+ G_{ijk}\epsilon_i\epsilon_j\epsilon_k + \ldots$
F's and G's are material Constants	

7.5. Laminate Strength Analysis

We now turn our attention from single lamina strength analysis to the failure behavior of multiply laminae or laminates. The objective of this analysis is to determine the strength behavior of each lamina in the laminate assuming that a plane state of stress exists for each lamina independent of its orientation and position within the laminate. The application of a suitable failure criterion must be adopted, the stress on each lamina calculated and transformed to the material axes of the lamina, and a fail safe measure of each lamina determined. Several points must therefore be addressed in terms of general laminate strength analysis. We begin by examining several approaches to laminate strength analysis which have been advanced.

In the broadest context of laminate strength interrogation, one is generally knowledgeable of either,

- Knowing the loads/Testing the design
- Knowing the design/Testing for allowable loads

For the first case indicated, so-called knowledge of the loads may be available, however, generally this is more in the realm of the designers province to determine by experience rather than by rigorous deterministic methods. In the second case, many times the design is fixed and knowledge of what loads the structural elements can safely withstand is the challenge. Each of these issues will be examined through examples in this section. An awareness of the semi-global failure analysis of a laminate raises the following issues, that is,

- First Ply Failure
- Behavior after First Ply Failure

First ply failure occurs when the given or calculated loads exceed the ply specified failure criterion. Should this test be met then it is still possible for the system to carry additional loads after first ply failure and several approaches to accounting for ply failure can be introduced. Among these are,

- Total Ply Discount
- Ply Failure Distribution

The first approach, that of total ply discount, implies that the ply still remains within the system as a volume entity but that it no longer carries any load. This represents a conservative approach to failure analysis. The second approach implies that some knowledge of the failure mechanisms of the ply are known. For example, if matrix failure in a particular ply is known to occur, then the transverse properties of that ply can be omitted. Alternatively, if fiber failure occurs within a given ply, then that ply can be treated as having zero stiffness for subsequent analysis.

In addition to examining the effects of ply failure within the laminate, it is sometimes equally advantageous to establish failure envelopes based upon various combinations of a given set of applied loads. The most widely used loading for these latter schemes appear to be that of membrane type loads. Both interrogation of ply failure and failure interaction envelopes will be discussed in the examples to follow. Before discussing these examples, however, it is appropriate to review the pertinent laminate equations, including generalizations to incorporate both temperature and humidity effects, required for analysis of composite failure. These equations will be useful to the reader in examining more complex failure situations where residual stresses or environmental conditions may be important as well as in establishing interactive failure envelopes. Considering a plane state of stress and referring to an arbitrary axes x and y with respect to the principal coordinate material directions, we can write the following stress-strain equations for the kth

lamina of a stacking sequence comprising N lamina,

$$
\begin{bmatrix} \sigma_x \\ \sigma_y \\ \sigma_{xy} \end{bmatrix}_k = \begin{bmatrix} \overline{Q}_{11} & \overline{Q}_{12} & \overline{Q}_{16} \\ \overline{Q}_{21} & \overline{Q}_{22} & \overline{Q}_{26} \\ \overline{Q}_{61} & \overline{Q}_{62} & \overline{Q}_{66} \end{bmatrix}_k \begin{bmatrix} \epsilon_x - \alpha_x \Delta T - \beta_x \Delta m \\ \epsilon_y - \alpha_y \Delta T - \beta_y \Delta m \\ \epsilon_{xy} - \dfrac{\alpha_{xy} \Delta T}{2} - \dfrac{\beta_{xy} \Delta m}{2} \end{bmatrix}_k
$$

The strains $\epsilon_x^{(k)}$, and $\epsilon_y^{(k)}$, and $\epsilon_{xy}^{(k)}$, can be rewritten in terms of the displacement components u, v, and w resulting in the following equation,

$$
\begin{bmatrix} \sigma_x \\ \sigma_y \\ \sigma_{xy} \end{bmatrix}_k = \begin{bmatrix} \\ \overline{Q}_{ij} \\ \\ \end{bmatrix}_k \begin{bmatrix} \epsilon_{x_0} + z\kappa_x - \alpha_x \Delta T - \beta_x \Delta m \\ \epsilon_{y_0} + z\kappa_y - \alpha_y \Delta T - \beta_y \Delta m \\ \epsilon_{xy_0} + z\kappa_{xy} - \dfrac{\alpha_{xy} \Delta T}{2} - \dfrac{\beta_{xy} \Delta m}{2} \end{bmatrix}_k
$$

The stresses for the kth lamina can be redefined in terms of the membrane stress resultants N_x, N_y, N_{xy} and bending stress resultants M_x, M_y, M_{xy}. Integrating the thickness of the lamina as shown in equation (2.47), we obtain equation 2.49. In matrix notation this can be written as,

$$ [N] = [A][\epsilon_0] + [B][\kappa] + [N]^T - [N]^m $$

A similar set of equations can be found for the bending stress resultants, that is

$$ [M] = [B][\epsilon_0] + [D][\kappa] - [M]^T - [M]^m $$

transposing $[N]^T$, $[N]^m$, $[M]^T$, and $[M]^m$ from the right hand side to the left hand side of the stress resultant equations we can write:

$$ [\overline{N}] = [N] + [N]^T + [N]^m = [A][\epsilon_0] + [B][\kappa] $$

$$ [\overline{M}] = [M] + [M]^T + [M]^m = [B][\epsilon_0] + [B][\kappa] $$

The barred quantities are often referred to in the literature as the total force and moment resultants.

In some cases, it is advantageous to have the mid-plane strains and curvatures defined in terms of the stress resultants. Thus, inverting the above equations gives:

$$
\begin{bmatrix} \epsilon_0 \\ \kappa \end{bmatrix} = \begin{bmatrix} a - ba^{-1}b^T & ba^{-1} \\ -d^{-1} & d^{-1} \end{bmatrix} \begin{bmatrix} \overline{N} \\ \overline{M} \end{bmatrix}
$$

Thus for the case of plane stress of anisotropic laminates, in the absence of transverse shear stress, the principal failure criteria can be interrogated as follows,

Maximum Stress Theory

(1) Establish the stresses acting on the system
(2) If the stresses do not coincide with the principal material direction they should be transformed according to the transformation law $[\sigma_L] = [T][\sigma_x]$
(3) The stresses $[\sigma_L]$ should then be compared with the allowable stresses on a ply by ply basis to establish whether first ply failure has occurred.
(4) The process of estimating whether laminate failure has occurred by introducing a selected post first ply failure criterion can then be evaluated.

Maximum Strain Theory

(1) Establish the stresses acting on the system
(2) Calculate the strains corresponding to the applied stresses
(3) If the calculated strains do not coincide with the principal material directions they should be transformed according to the transformation law
(4) The strains $[\epsilon_L]$ should then be compared with the allowable strains on a ply by ply basis in order to establish whether first ply failure has occurred.
(5) The process of estimating whether laminate failure has occurred by introducing a selected post first ply failure criterion can then be evaluated.

Interactive Failure Criteria

(1) Establish the stresses acting on the system.
(2) If the stresses do not coincide with the principal material directions they should be rotated using the transformation $[\epsilon_L] = [T][\epsilon_x]$
(3) The stress $[\sigma_L]$ should then be inserted into the selected interactive failure criterion, that is, Tsai-Hill, Tsai-Wu, Hoffman, and the criterion tested to ensure existence of a fail safe design on a ply by ply basis.
(4) If a failed ply has been detected, an appropriate first ply failure assessment can be made and the composite system interrogated for laminate failure.

7.6. Problems

7.1. Consider the following laminate of equal ply thickness 0.0052″

$$\begin{vmatrix} \theta = 30° \\ \theta = 0° \\ \theta = -30° \end{vmatrix}$$

Assuming the maximum strain theory governs failure, with the following inequalities given

$$|\epsilon_1| \leqslant 0.004$$

$$|\epsilon_2| \leqslant 0.003$$

$$|\sigma_{12}| \leqslant 0.010$$

Determine whether failure will occur at $z = 0$ for the following loadings.

$$N_x = 100 \text{ lb/in} \quad M_x = 5 \text{ in-lb/in}$$

The properties of an unidirectional layer are given by:

$$E_{11} = 29.2 \times 10^6$$

$$E_{22} = 2.4 \times 10^6$$

$$\nu_{12} = 0.223$$

$$\nu_{21} = 0.018$$

$$G_{12} = 0.83 \times 10^6$$

7.2. Consider the pure shear loading of a composite material at any angle θ with respect to the principal material directions. Use the Tsai-Hill criterion to find failure of the composite for this loading.

7.7. References

1. E.M. Wu, "Mechanics of Composite Materials" in Composite Materials, L.T. Brout-man and R.H. Krock (Eds.) Academic Press, Vol. 2, p. 354, 1974.
2. W.J.M. Rankine, *Applied Mechanics*, 1st edition, London, 1858.

238

3. B. De. St. Venant, "Memoire sur les equations generales des mouvements interieurs des corpes solides ductites an dela des limites ou l'elasticite pourrait des ramener a leur premier etat" Comptes Rendus, Paris, France, Vol. 70, pp. 473–480, 1870.

4. H. Tresca, "Memoire su l'ecoulement des corps solides soumis a de fortes pressions", Comptes Rendus, Paris France, Vol. 59, p. 754, 1864.

5. R. Von Mises, "Mechanik der festen Koerper im plastisch - deformablen zustand", Goettinger Nachrichten. Mathematisch, Physikalische Klasse, Vol. 1913, pp. 204–218, 1913.

6. R.S. Sandhu, "A Survey of Failure Theories of Isotropic and Anisotropic Material", AFFDL-TR-72-71, Wright Patterson Air Force Base, Dayton, Ohio, 1972.

7. B.E. Kaminski and R.B. Lantz, "Strength Theories of Failure For Anisotropic Materials", Composite Materials: Testing and design ASTM 460, pp. 160–169, 1969.

8. E.Z. Stowell and T.S. Liu, "On the Mechanical Behavior of Fiber Reinforced Crystalline Solids", Journal of the Mechanics and Physics of Solid, Vol. 9, pp. 242–260, 1961.

9. P.W. Kelly, G.J. Davies, Metallurgical Reviews, Vol. 10, 1965.

10. M.E. Waddoups, Advanced Composite Material Mechanics for the Design and Stress Analyst, General Dynamics, Ft. Worth, TX, Report FZM-4763, 1967.

11. J. Marin, Theories of Strength for Combined Stresses and Non Isotropic Materials, J. of Aerospace Sciences, Vol. 24, pp. 265–268, 1957.

12. C.B. Norris, Strength of Orthotropic Materials Subjected to Combined stress, Forest Products Laboratory Report 1816, 1950, 1962.

13. V.D. Azzi and S.W. Tsai, "Anisotropic Strength of Components" Experimental Mechanics, Vol. 5, pp. 286–288, 1965.

14. O. Hoffman, "The Brittle Strength of Orthotropic Materials", J. Composite Materials, Vol. 1, pp. 200–206, 1967.

15. S.W. Tsai and E.M. Wu, "A General Theory of Strength for Anisotropic Materials", *Jl. Composite Materials*, Vol. 5, pp. 58–80, 1971.

16. R.C. Burk, "Standard Failure Criteria Needed for Advanced Composites," Astronautics and Aeronautics, Vol. 21, pp. 58–62, 1983.

8. Joining of Composite Material Structures

8.1. General Remarks

In the design of composite material structures, components must be joined in such a manner that the overall structure retains its structural integrity while performing its intended function, subjected to loads (static and dynamic) and environment (temperature, humidity).

The use of composite materials in complex structures rather than metals almost always significantly reduces the number of components, sometimes from hundreds to dozens, with the advantages of great savings in weight, cost, inspection, assembly, and increased reliability. Yet, joining is still required. Joining metallic structures is a well developed technology involving riveting, bolting, welding, glueing, brazing, soldering, and combinations thereof. However, for polymer matrix fiber reinforced composites only adhesive bonding and mechanical fasteners (bolts and rivets) can be utilized.

Inherently, adhesive bonding is preferred because of the continuous connection, whereas in drilling the holes for the bolts or rivets, fibers are cut, the joining is at discrete points, and large stress concentrations occur around each hole drilled. However, in many structures, some structural components must be removed periodically for access to the interior (for example access to electronic components in an aircraft, etc.) so invariably some use of mechanical fasteners must be made. In this chapter both adhesive bonding and mechanical fastening are discussed and described in detail. This important area of technology is growing and changing very rapidly and, as will be seen, much more effort is needed to solve important problems particularly in the adhesive bonding area.

8.2. Adhesive bonding

8.2.1. Introduction

Adhesive bonding for joining materials began when the Egyptians used copper chloride poisoned casein adhesives in fabricating mummy

Table 8.1. Joining Methods Utilization

Methods	1963/65	1966/68	1969/71	1972/74	1975/77
Adhesives	$458 MM	$560 MM	$669 MM	$1,170 MM	$1,655 MM
Bolts, Nuts					
Fasteners	1,204	1,663	1,688	2,400	2,840
Nails & Spikes	161	175	191	278	326
Welding Materials					
Solder, etc.	366	400	455	537	699
Total	$2,189 MM	$2,798 MM	$3,003 MM	$4,385 MM	$5,520 MM
% Adhesives	20.9%	20.0%	22.3%	26.7%	30.0%

cases. However, adhesive bonding for load bearing structures developed only several decades ago. One could say that the bonding of the plywood laminae of the British wooden Mosquito bomber of World War II might be the first aviation application. In the United States, Narmco developed the Metlbond adhesives in the early 1940's for the Consolidated Vultee B-36 bomber. Since that time the use of adhesive bonding has increased steadily. Kuno gives some statistics in Table 8.1.

The reasons why adhesive bonding in both metallic and composite material structures is so desirable compared to other joining methods are manifold according to Kuno [1]; they include:

1. Often, thinner gauge materials can be used with attendant weight savings and cost savings, for example the case of adhesive bonding aluminum sheets of 0.508 mm (0.020″) thickness, where 1.30 mm (0.051″) thick sheets would be required for riveting.
2. Number of production parts can be reduced, and the design simplified.
3. Need for milling, machining and forming operation of details is reduced.
4. Large area bonds can be made with a minimum work force without special skills.
5. Adhesive bonding provides a high strength to weight ratio with three times the shearing force of riveted or spot welded joints.
6. Aerodynamic smoothness and improved visual appearance.
7. Use as a seal, and/or corrosion preventer when joining incompatible adherends.
8. Excellent electrical and thermal insulation.
9. Superior fatigue resistance. Adhesive assemblies have shown fatigue life times twenty times better than riveted or spot-welded structures of identical parts.
10. Damping characteristics and noise reduction are superior to riveted or spot welded assemblies.
11. Often, the adhesive is sufficiently flexible to allow for the variations

in coefficients of thermal expansion when joining dissimilar materials.

Therefore, adhesive bonding is very desirable for use in composite material structures, and its use will be very beneficial from a weight, cost, reliability and flexibility of use. However, it is imperative that the ability to analyze, design and optimize these joints, of many configurations, subjected to generalized mechanical, thermal, and hygroscopic loads be further developed.

8.2.2. Review of Adhesively Bonded Joint Technology.

The following review provides a history and background of the present state of knowledge in adhesively bonded joints. It certainly is not all inclusive, but provides a basis to study other literature.

8.2.2.1. Single Lap Joint

The single lap joint, shown in Figure 1, has been studied more extensively than any other configuration through many analytical, finite difference and finite element methods. The figure shows a cross-section of two panels, referred to as adherends, joined together by the adhesive.

Early methods of analysis for isotropic (metallic) adherends include those of Szepe [2], Volkersen [3], De Bruyne [4], Wang [53], and Goland and Reissner [5]. These are discussed in detail by Kutscha and Hofer [6], of Illinois Institute of Technology, who performed an excellent review of all work in this area from 1961 to 1969. Work prior to that date was reviewed by Kutscha [46]. They state that the literature is very meager. Under AFML Sponsorship they conducted both an analytical and experimental study to determine the feasibility of developing rational design techniques for several bonded joint configurations, including the single lap joint in advanced composite structures, using the methods developed by the authors named above. Kutscha and Hofer conclude in 1969 that the present state of design in all bonded joints is simply that no rational procedure exists, and that such an approach must await further developments both in analytical methods and in materials characterization. Until that time design techniques will remain empirical and thus rely heavily on testing and experimental results. Since that time, more

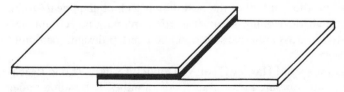

Fig. 8.1. Single Lap Joint

comprehensive methods of analysis have been developed which counter the above statement to some extent.

Kutscha and Hofer [6] found that in a single lap joint with adherends of unequal thickness, the maximum shear stress will occur at "the point where the load enters the joint from the thinner adherend". In a joint with equal thickness adherends the stress is the same at both ends of the overlap. Today the same conclusions are found, except that it is known that maximum shear stresses occur near to the end not at the end of the overlap.

Kutscha and Hofer's parametric studies show several important generalizations. Adhesive stresses decrease with increasing specimen width, up to about a four inch width, beyond which the stresses remain constant. Thus, wider joints designed on one inch wide data should be conservative because wider joints should be stronger than predicted. Also, the maximum stresses do not decrease significantly as the bonded area increases. As adhesive film thickness increases, more adhesive material is available to absorb differential straining and stresses decrease. As shear modulus increases, the maximum stresses increase almost linearly with the shear modulus for values between 50,000 psi to 250,000 psi, the conventional range of values for structural adhesives. They also find that as the stiffness of the adherends increase, the resistance of the joint to bending increases, and therefore the maximum stresses decrease. Lastly, maximum stresses in the adhesive are relatively insensitive to the value of the Poisson's ratio of the adherends. All of these early conclusions and generalizations hold today. Kutscha and Hofer also evaluated the following adhesives: FM 1000; FM 47, Type 2; Metlbond 040C; AF-131; Epon 422; and E 787.

Kutscha and Hofer also define a "joint efficiency," that is very useful. It is defined for any bonded joint as the axial load divided by the nominal bonded area, divided by the strength of the weaker adherend without the joint.

It is interesting to note that the fatigue tests conducted were at 15 cycles per second, a frequency low enough to prevent significant heating in the viscoelastic adhesive. They defined fatigue runout to be 10^7 cycles, that is fatigue runout is the number of cycles at which testing is terminated if no fracture occurs prior to that time. In stress calculations they assumed the Poisson's ratio of all of the adhesives to be 0.45.

In conclusion, they state that the best design technique appears to be empirical through developing shear strength joint parameter data, use this as a guide for gross adhesive selection and overlap design, then build a joint and test it.

In 1969, Lehman and Hawley [7] of McDonnell Douglas (Long Beach) also analyzed and compared several methods of adhesive bonding under AFFDL sponsorship, including single lap, double lap, scarf, stepped lap and variable adhesive joint configurations using a lumped parameter model. Their goal was to determine both static and fatigue strengths of these various configurations. Although transverse shear deformation, transverse normal strain, hygrothermal and viscoelastic effects were not considered, good agreement was found between analytical prediction and experimental results. However, their methods did include non-linear effects, effects which were first deemed to be important by Hartman and de Jonge [5] in 1960. Hartman and de Jonge also suggested the application of fracture mechanics to bonded joint analysis. Lehman and Hawley also found that the maximum joint strength occurs when the extensional stiffness of both adherends is the same. Also, they found that fatigue runout occurs when the maximum shear stress in the adhesive is below the proportional limit shear stress in the adhesive. Adhesive shear stresses above the proportional limit caused fatigue cracks in the adhesive that propagated through the joint causing failure. They also found that the residual strength of the fatigue specimens that survived runout usually exceeded static strength values. They also found that in bonded joints of epoxy resin composites with L/t ratios of about 25, interlaminar shear strength of the laminate is the limiting strength. They also provided a good data base of laminate and adhesive properties, thus helping to correct a deficiency stated by Kutscha and Hofer [6]. Lehman and Hawley [7] also conclude that semi-empirical methods are the most effective approach to rational joint design.

Earlier, in 1960 Wang [53] had shown that using bonded joints employing adhesives with lower moduli resulted in better fatigue strength.

In 1972, Dickson, Hsu and McKinney [29], of Lockheed Georgia under AFFDL Sponsorship, developed the BONJO 1 linear analysis program for bonded joints in laminated composites. They compared their results with finite element analyses, strain gage data and photoelastic data. The BONJO 1 program was extended to include perfectly elastic-plastic adhesive stress-strain behavior. Dickson also discusses the deleterious effects of residual thermal stresses resulting from cool down after curing. He also comments that the use of large finite element programs, which are capable of virtually any degree of complexity, are both cumbersome, costly and time consuming – noting that it is preferable to use closed form solution methods whenever possible. They analyzed single lap joints, the double lap joint, and both the single doubler and double doubler joints, including plasticity effects and also included transverse shear deformation and transverse normal deformation. Also in 1972, Grimes and his colleagues at Southwest Research Institute [10] under AFFDL sponsorship utilized a discrete element method and a continuum elasticity method to predict adherend stresses and joint strengths. Material non-linear effects in adherends and adhesives were included in their study of single, double and step lap configurations. Experimental verification was also accomplished. In their analysis, the adhesive transverse normal and shear stresses obey a Ramberg-Osgood [56] approximation. Experimental verifications of the two methods of analysis were made. Accurate predictions of adherend stresses and joint strength was demonstrated for static loadings; however, Grimes notes that transverse shear stress, transverse normal strain, hygrothermal considerations and viscoelastic effects although neglected should be included. They also point out that small deflection theory is not valid for the continuum analysis due to the large deflections observed in joints under load. In predicting failure loads in the experimental program, Grimes used maximum stress theory for the adhesive and for isotropic adherends; maximum strain theory was used for composite adherends. They also note the much larger amount of computer time required by the discrete element solution compared to the continuum solution. Grimes also added to the data base for both adherends and adhesive properties.

For design, Grimes states that since the predicted mean strength and type of failure or even complex joints is reasonably accurate when compared with experimental results (203 tests), design allowable calculated from mean strength predictions should also be accurate. Thus, the use of the standard 1.5 factor of safety on design limit loads to obtain design ultimate loads should be sufficient for static loadings at room temperature.

The work of Hart-Smith [8,9] on single lap joints in 1973 is also notable. The analysis, sponsored by NASA-Langley, involves a continuum model in which the adherends are elastic and the adhesive bond is elastic-plastic in shear while behaving elastically in transverse tension. Allowance is made for the peel (tensile) stresses at the ends of the adhesive, temperature differences between the adherends and differences in adherend stiffness. Comparisons in analysis are made to the earlier work of Volkersen [3], and Goland and Reissner [5]. In fact, Oplinger [25] notes that Hart-Smith utilizes the Volkersen membrane response approach for the adherends. No experimental verification of the analysis was undertaken. Transverse shear deformation, and transverse normal strain are not considered. Hart-Smith concludes that the influence of the adhesive on joint strength is determined by the shear strain energy to failure of the adhesive bond.

Hart-Smith [9] finds that yielding in the metal adherends at the ends of a single-lap joint usually initiates failure for all but very short overlap lengths. For composite adherends, fracture of 0° filaments close to the adhesives usually initiates an interlaminar shear failure within the laminates. He also finds that for thicker adherends, the dominant failure modes are peel tensile stresses in the adhesive and the associated interlaminar tension stresses in the composite adherends.

He also points out that any adherend imbalance causes a significant strength reduction in a bonded joint, even compared to the strength of a balanced joint employing only the geometry of the weaker adherend of the unbalanced joint.

As a design suggestion, because the eccentric load path of a single lap joint results in a low structural efficiency, a larger L/t ratio should be employed than that needed for load transfer alone in order to increase structural efficiency even though a small weight penalty results. At one point he states that single-lap joints are so inefficient however that they should not be used without some support to react the eccentricity.

To characterize the stresses in the adhesive, Hart-Smith treated the transverse tension, σ_{adh}, as

$$\sigma_{adh} = E_{adh} \frac{w_3 - w_2}{\eta} \tag{8.1}$$

where E_{adh} is the adhesive tensile modulus, w_2 and w_3 are the lateral deformations on the adherends above and below the adhesive layer, and η is the adhesive thickness.

For the elastic shear behavior of the adhesive, τ_{adh}, the equation is

$$\tau_{adh} = G_{adh} \frac{u_3 - u_2}{\eta} \quad \text{for} \quad \gamma \leqslant \gamma_e \tag{8.2}$$

where G_{adh} is the shear modulus of the adhesive, and U_2 and U_3 are the in-plane displacements of the adherends above and below the adhesive layer, and γ is the "engineering" shear strain. Then, Hart-Smith modelled the adhesive shear stress to be elastic-perfectly plastic, so,

$$\tau_{adh} = \tau_p \quad \text{for} \quad \gamma \geqslant \gamma_e \tag{8.3}$$

where τ_p is the adhesive shear failure stress, for shear strains above the elastic limit γ_e.

The rationale Hart-Smith uses for neglecting plasticity in tension in the adhesive is twofold. First, he asserts that for composite adherends, the adherend is much weaker in transverse tension than is the adhesive, hence it will fail before any plastic deformation of the adhesive will occur. Secondly, the adhesives, being long-chain polymers, are essentially incompressible, and hence the adherend constraints on the adhesive suppress plastic deformation in tension or compression.

Hart-Smith also notes that when adherends have differing coefficients of thermal expansion, and when the joint operates at a temperature differing from the cure temperature, the joint load capacity usually is adversely affected. He notes (1973) that analytically an iterative technique must be employed because explicit expressions cannot be obtained.

Hart-Smith's analysis establishes also that tough ductile adhesives produce much stronger joints than those which are stronger but more brittle.

Oplinger [22] comments that the Hart-Smith approach provides important insight into the role played by adhesive plasticity. However, he questions the consistency of assumptions because of the elastic-zone peeling stress assumption and the elastic-perfectly plastic behavior of the assumed shear stresses. He states that Dickson and co-workers [29] and Grimes and co-workers [10] appear to employ a more consistent approach utilizing an effective stress and a Ramberg-Osgood Law [56] for plastic strain behavior.

Commencing 1973, a number of publications by Renton and Vinson [11–18] and Renton, Vinson and Flaggs [19], University of Delaware under AFOSR Sponsorship dealt with analytical methods for analyzing stresses, strains and displacements in both the adhesive bond and adherends in single lap joints under in-plane static and dynamic loads. Transverse shear deformation, and transverse normal strain were included, as well as temperature effects in some of the publications referenced above. The methods of analysis were assumed to be linearly elastic only because it was generally agreed that for fatigue runout to occur, the maximum stresses in the adhesive must remain below the adhesive proportional limits in both tension and shear.

In 1974, Renton and Vinson reported on the results of a series of fatigue tests using a ductile adhesive, Hysol EA 951, and two different anisotropic adherend materials, subjected to constant amplitude and a two-block repeated load spectrum. The results showed that the proportional limit stress of the adhesive is a very important parameter. Fifty two unidirectional adherend specimens and fifty one angle ply fiberglass epoxy specimens were tested as well as nineteen Kevlar-49-epoxy specimens. It was found that the Kevlar-49 specimens were superior in performance. They also found that a 20–40% reduction in attainable loads associated with runout occurs when the angle ply construction above is used compared with the unidirectional construction. Also in 4×10^6 cycles, a maximum design load corresponding to 26% of ultimate static load was attained for a lap length of 0.30 inches, and 20% for a lap length of 0.60 inches independent of ply orientation. No adhesive thickness effects on fatigue life were identified.

Early support for the Renton-Vinson methods of analysis, codified as BOND 3 and BOND 4 for single lap bonded joints, came from Sharpe and Muha [57]. They measured the shear strains in the adhesive layer of a single lap joint by monitoring the fringe pattern generated by a laser incident on imbedded single wires distributed along the adhesive layer. From this they calculated the shear stresses. They chose the single lap joint "because there are significant normal stresses present and it is

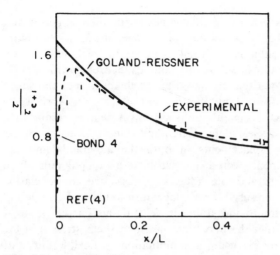

Fig. 8.2.

presumed that the theory that can best predict the measurable shear stresses under these complicated conditions will be the best theory." Comparing their experimental results with over twenty finite-element and closed form analytical solutions, they found the Renton-Vinson closed

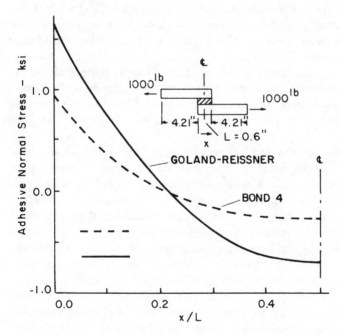

Fig. 8.3.

form solution of BOND 4 to agree the best with their experimental results. Figure 8.2 explicitly shows the comparisons made, while Figure 8.3 illustrates typical adhesive normal stresses. In each case corresponding results of the Goland-Reissner methods [5] are shown.

In 1977, Sen [33] stated that the most complete analysis of the single lap joint has been performed by Renton and Vinson. In April 1975, Srinivas [21] of NASA-Langley published his research results. He accounted for transverse shear deformation and transverse normal stress to obtain stress and displacements in both the adherends and the adhesive. The displacements were expanded in terms of polynomials in the thickness coordinate and the adhesive is assumed to be elastic. Equations (1) and (2) were used. Single-lap, flush joints and double lap joints were analyzed. He provides a multitude of design curves. Some of the generalizations are that the maximum peel and shear stresses in the bond can be reduced by (1) using a combination of flexible and stiff bonds, (2) using stiffer lap plates, and (3) tapering the plates. Of the three joints above the double lap joints had the smallest values of adhesive stresses while the flush joint had the highest maxima.

In 1975, Oplinger [22] published the results of his research. Oplinger's classical paper went far to organize the plethora of publications that existed to that date. He compared and evaluated several analytic and finite element models of bonded joints, and concludes that transverse shear deformations, transverse normal strain temperature effect, non-linear adherend and adhesive behavior, viscoelastic adhesive behavior and fracture mechanisms need to be included in modelling a bonded joint of any configuration.

Oplinger uses the concept of an ineffective length, defined as the lap length beyond which an increase in that dimension is ineffective in reducing peak adhesive shear and peel stresses. Equations are given for that length and those peak stresses, but are too involved to be given herein.

In 1976, Wetherhold and Vinson [23] generalized the Renton-Vinson model to include hygrothermal effects in the adherend. This model can be described very simply. Figure 8.4 represents a model for a structural member in the x-z plane, assuming plane strain in the y-direction, characteristic of a wide plate or shell structure loaded in the x-z plane. A closed form analytic solution was developed for this structural element subjected to stress resultants N_i, stress couples M_i, and shear resultants Q_i. The structure is also subjected to distributed normal loads $p_u(x)$ and $p_1(x)$, and distributed shear loads $\tau_u(x)$ and $\tau_1(x)$ as shown. Hygrothermal effects (effects of combined high temperature and high humidity) are included. The solutions are general for most practical stacking sequences, but are restricted to midplane symmetry, and (as stated before) to plane strain in the y direction. Thus, this solution becomes a finite element

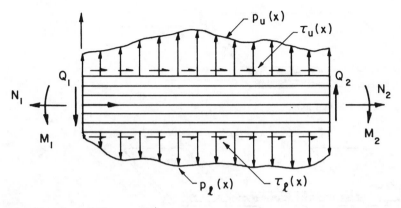

Fig. 8.4. Adherend Element

type solution (and is analytic) usable as a building block for any adhesive joint configuration, such as a single lap joint as shown in Figure 8.1 where each of the adherend elements is a special solution of Figure 8.4 with appropriate loads and boundary conditions. This concept has proven to be most useful. The adhesive is modelled by Equations 1 and 2, as done by several other researchers.

During 1976 and 1977, additional research on single lap bonded joints was published by Liu [24], Allman [25], and Humphreys and Herakovich [26]. Allman restricts his solutions to linear elastic adhesive behavior because certain epoxy-resin adhesives have a linear stress-strain relation to failure. As so many before, he uses a plane-strain assumption. Humphreys and Herakovich, also using a plane-strain finite element model, assumed a Ramberg-Osgood Stress-Strain Model for both the adhesive and adherend. Alteration in material properties due to hygro-thermal exposure was included through using percent retention curves.

In 1979, Vinson and Zumsteg [27], under ONR sponsorship, corrected some small errors in the Wetherhold-Vinson analysis [23], but more importantly found a unifying method, using the building block approach depicted by Figures 8.1 and 8.4 to analyze many single lap joints, double lap joints, single doubler joints and double doubler joints (Figures 8.5 through 8.7 respectively) using one analysis, and merely modifying a few boundary conditions and adherend thicknesses. This reduces the analytical effort to a minimum.

Most recently, Flaggs and Crossman [28] used a linear hereditary integral viscoelastic constitutive law for the analysis of composite adherend single lap joints under mechanical loading and/or transient hygrothermal exposure. The viscoelastic effects are significant. The Flaggs-Crossman approach allows the analysis of stresses in bonded joints over periods of years, utilizing creep relaxation data from experi-

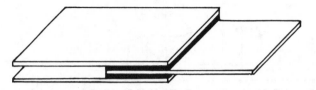

Fig. 8.5. Double Lap Joint

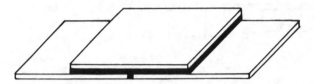

Fig. 8.6. Single Doubler Joint

ments performed at various temperatures and moisture contents performed over periods of several minutes.

8.2.2.2. Double Lap Joint and Double Doubler Joint

The double lap joint has also been studied by many researchers. There, however, is an ambiguity in the definition of the double lap joint Figure 8.5, some calling Figure 8.7 a double lap joint.

Methods of analysis developed for one or both of these configurations include those of an AVCO group [51], Kutscha and Hofer [6], Lehman and Hawley [7], Dickson [29], Grimes [10], Hart-Smith [30,31], Srinivas [21], Keer and Chantaramungkorn [32], Oplinger [22], Allman [25], Sen [33] and Humphreys and Herakovich [26]. The AVCO group employed a finite element technique to study the behavior of a unidirectional boron composite bonded to two panels of titanium. They note a major discrepancy with the earlier Volkersen [3] analysis. Lehman and Hawley [7] were the first to show that double lap joints were more than twice as strong as single lap joints of the same lap length due to the symmetry of this configuration, which reduced bending in the adherends and transverse stresses across the adhesive. They also modified their linear analysis to account for plasticity in an approximate manner, which agreed well

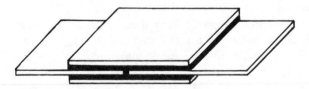

Fig. 8.7. Double Doubler Joint

with experimental data. Early on (1965), the Russian, Prokhorov [54], using previously developed methods of analysis, suggested that for practical design of double lap joints, a lap length to adherend thickness ratio of thirty be used. Keer used integral transform techniques to study the double lap joint. Sen [33], 1977, considered a linear viscoelastic adhesive and the SAAS III finite element program to study the double lap joint.

Again, the Vinson-Zumsteg [27] methods can be employed to analyze double lap and double doubler joints, with all of the comprehensive effects included.

8.2.2.3. Single Doubler Joint

Relatively little work has been performed on this configuration, see Figure 8.6. However, it could be of importance in some repair procedures. Because of the asymmetry it results in high peel stresses, analogous to the single lap joint.

Again, through the building block method of analysis depicted in the Vinson-Zumsteg [27] paper rational methods of analysis, design and optimization can be developed which encompass all effects important to bonded joints.

8.2.2.4. Scarf Joints

The scarf joint is depicted in Figure 8.8 below.

Its advantage is in aerodynamic smoothness, but the disadvantage is in the careful machining required to have a uniform bond line. Therefore, it is perhaps more useful for metallic adherends rather than those composed of composite materials. Only a few researchers have studied scarf joints. Analysis methods available include those of Kutscha and Hofer [6], Lehman and Hawley [7], and Oplinger [22]. Lehman and Hawley [7] found that the scarf joint and the stepped lap configurations have the capability of scale-up to transmit a load of any magnitude providing that the lap length is not restricted. They also found that in fatigue ($R = +0.05$, tension) scarf joints sustained stress levels 3.5 times greater than those of double lap joints. They found that the scarf joint approaches the ideals of strain compatibility in the adherends and uniform stress in the adhesive. One result of this is that adhesive ductility in scarf joints is less important than in other configurations.

Fig. 8.8. Scarf Joint

Grimes [10] suggests in 1973 that a nonlinear analysis for scarf joints is needed. Oplinger [25] states that both scarf and step lap joints provide efficient use of the full overlap length by allowing the peak adhesive stresses to be reduced as the overlap length increases, thus avoiding the "ineffective length' bound, defined earlier.

Analysis of scarf and step-lap joints also include closed form solutions by Hart-Smith [58], Erdogan and Ratwani [59] along with a number of finite element approaches [29,60]. In particular, comparisons of finite element results with closed form analytical results [59] are discussed by Barker and Hatt [61].

8.2.2.5. Stepped Lap Joint

The stepped lap joint is shown in Figure 8.9.

Methods of analysis available for one or both of these configurations have been formulated by Kutscha and Hofer [6], Lehman and Hawley [7], Grimes [10], and Oplinger [22]. Lehman and Hawley found that stepped lap joints achieve higher average shear stresses than the scarf joint; also the strength of the stepped lap joint is not sensitive to the number of steps when the total lap length is held constant. In fact, they found that scarf and stepped lap joints are lighter in weight than other lap joints at all load levels. The stepped lap joint approximates the strain compatibility of the scarf joint. Static strength was not sensitive to the number of steps. In Grimes' study [10] he assumes that no force is transmitted by the adhesive in the riser of the step-lap configuration.

It should be noted that as the number of steps increase the stepped lap joint approaches the scarf joint configuration. So, the building block approach of Figure 8.4 could also be used to design and optimize the scarf joint.

8.2.2.6. Other Joint Configurations

Both Kutscha and Hofer [6] and Lehman and Hawley [7] discuss the dual adhesive or variable adhesive joint. This concept utilizes a very strong adhesive where the shear and tensile stresses peak, while using a weaker but perhaps more ductile adhesive in the lower stress region. Neither of the authors above feel the dual or variable adhesive joint concept has merit. It therefore is noted but will not be discussed further.

Fig. 8.9. Stepped Lap Joint

The concept of bonding and also employing mechanical fasteners has been studied by Lehman and Hawley [7] and Hart-Smith [29]. Lehman and Hawley found that the combination of bonding and mechanical fasteners results in better joints than found with bonding alone or mechanical fasteners alone, due to changes in failure modes with the combination.

The use of a smooth joggle to reinforce a single lap joint was suggested by Hart-Smith [9], but no tests were apparently conducted. (See Figure 25, Reference 9).

Tapered adherends and tapered doublers in several configurations, as well as notched adherends have been studied by Renton, Pajerowski and Vinson [17]. Some of them appear to be superior to the more classical configurations discussed earlier. More research is needed to confirm the early findings.

An interesting tapered joint configuration which shows promise was studied recently by Cooper and Sawyer [64].

8.2.2.6. Present Computer codes for Joint Analysis

All of the methods of analysis that have been developed for the analysis of adhesively bonded joints are very involved, and can best be used for design and optimization by employing computer codes.

The Aerospace Structures Information and Analysis Center (ASIAC) of the Air Force Flight Dynamics Laboratory (AFFDL) has the following programs available as of June 1978 for those with a valid need. They include the following:

BONJO, BONJOIG, BONJOIS – These analyze narrow, uniaxially loaded bonded lap joints. The adherends may be either isotropic or a laminated anisotropic panel. Effects of interlaminar shear and normal stresses, as well as residual stresses caused by bonding at elevated temperatures. BONJOIG analyzes any general single lap or double lap joint. BONJOIS analyzes any single lap bonded joints with identical adherends. BONJO approximates nonlinear properties through a linear elastic-perfectly plastic stress strain relationship. The code is based upon analysis of AFFDL TM 74-187.

JTSDL – This program performs a nonlinear analysis of single or double lap bonded joint in plane strain subjected to static loads at room temperature. Adherends can be orthotropic symmetric laminates or an isotropic material of constant thickness, where transverse normal stress and interlaminar shear are neglected. The adhesive is assumed to be isotropic and of constant thickness. The analysis is available in AFFDL-TM-73-3-FBC, which also covers JTSTP below.

JTSTP – This program is similar to JTSDL above but analyzes step lap joints of arbitrary number and geometry.

A4EH – This program analyzes double lap joints.

A4EN – This program analyzes single lap joints.

A4ES – This program analyzes scarf-joints.

The analyses for the last four programs above is given in AFFDL-TR-78-38.

8.2.2.7. Adhesive Material Property Characterization

In order to analyze, design and optimize a bonded joint of any configuration, it is necessary to know the tensile and shear moduli, proportional limit stresses, ultimate stresses, and strains to failure. This was recognized by Kutscha and Hofer [6] in 1969 who measured the static strength of three adhesives bonded to fiberglas, steel and titanium, as well as the fatigue strength. Prior to them, the research of Kuenzi and Stevens [47], Forest Products Laboratory is paramount. They characterized the adhesive shear properties using a thin walled torsion specimen. Their work was extended by Rutherford [48] who developed transducers for the torsion test accurate to 6.1×10^{-8} radians. He also used a butt joint specimen to obtain adhesive tension and compression properties using a capacitance type extensometer accurate to 1.27×10^{-5} mm (5×10^{-7} in), which for a bond line thickness of 0.127 mm (0.005 in) would correspond to measuring a strain of 1×10^{-8}.

Kutscha and Hofer criticize Rutherford's work only in the data reduction of the tensile butt joint tests, because they assumed that the stress state in the adhesive film was uniaxial, hence that the modulus of elasticity was simply load divided by cross-sectional area. Actually a three dimensional stress state exists. Hence Rutherford's conclusions regarding tension properties is inconclusive.

Also in 1969, Lehman and Hawley [7] characterized six adhesives including AF130 & Shell 951, using 12.7 mm (1/2") double lap specimens, flatwise tension tests on sandwich specimens, and in a torsional ring fixture they designed during their program.

Furthermore, in 1967, Franzblau and Rutherford [48] found that the elastic modulus of an adhesive in very thin film form was approximately twice that of the modulus of the bulk form, thus establishing the need for developing film tests to measure properties. The primary "standard" for adhesive mechanical properties that exists is one for shear properties, the ASTM Test for Strength Properties of Adhesives in Shear by Tension Loading (Metal to Metal), (D1002-72). The ASTM specimen is a single lap joint, see figure 8.10 comprised of two thin adherends of representative joint materials, usually steel or aluminum, bonded together by an adhesive layer. The specimen is pulled in tension, and the mean value of shear strength is computed to be the maximum tensile load divided by the shear area.

Because of the thin adherends, the tensile in-plane load causes considerable bending in the joint, as seen in Figure 8.10.

Fig. 8.10. Single Lap Joint in Tension

This results in stress profiles in the adhesive shown typically in figures 8.2 and 8.3. It is seen that the shear stress peaks near to the ends of the single lap joint and of course goes to zero at the ends because it is a free surface. It is also seen that the normal or "peel" stress peak at the ends, while they are compressive over a sizeable region in the joint interior. The normal stresses can be even larger than the shear stresses in one joint.

It is seen that in the ASTM test the adhesive shear stress is far from constant, thus negating any value of adhesive shear strength calculated by computing the ratio of maximum tensile load divided by the shear area. Even worse is the fact that the maximum tensile or peel stress in the adhesive can be greater than the maximum shear stress. Hence, for many adhesives in which the ultimate tensile strength is approximately equal to the ultimate shear strength, failure of the adhesive in this "shear test" will initiate from exceeding the tensile strength, and it is no shear strength at all.

Hart-Smith [9] criticized the ASTM D-1002 tests in 1973, and states lucidly that the standard tests have been performed in the mistaken belief that the failure load was related to adhesive shear strength, and then proceeds to "debunk that myth." One explicit criticism is because the joint strengths do not scale linearly with bond area.

In 1969, Lehman and Hawley [7] used a torsion ring shear test specimen which they said eliminated large stress concentrations and secondary peeling effects in lap joint specimens. However, there are many disadvantages in using this specimen. However, they found that the ultimate shear stress and strain and the elastic modulus decreased with adhesive thickness. Also they found that ultimate shear stress and elastic shear modulus increased and ultimate shear strain decreased with increasing strain rate.

Guess, Allred and Gerstle [34] also commented on the ASTM Test D1002-72 lap shear specimen for comparing relative strengths of different adhesives.

A better specimen to use is a single lap joint specimen as shown in Figure 8.11, wherein the flexural and extensional stiffness of the identical adherends are very, very large, compared to those of the Standard ASTM Specimen, as discussed by Renton and Vinson [14]. If they are sufficiently large, the adhesive stresses will be shown in Figure 8.11.

Renton and Vinson [11,13] analyzed various thick adherend specimens and concluded that if the configuration has a ratio of $A_{11}/G_a\eta > 25 \times 10^6$,

256

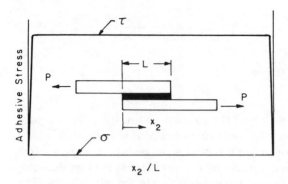

Fig. 8.11. Thick Adherend Test Piece and Stress Distributions

where A_{11} is the axial extensional stiffness of the adherend, G_a the expected shear modulus of the adhesive, and η the bond thickness, then the adhesive shear stress, $\tau(x)$, will be constant except of course at the free edges. Then, the shear strength is indeed load divided by area. Also if the configuration has a ratio of $D_{11}/E_a\eta^3 > 14.5 \times 10^9$, (where D_{11} is the axial flexural stiffness of the adherend, and E_a is the expected tensile modulus of the adhesive) then tensile stresses in the adhesive are negligible. This means that adherends in the order of 6.35 mm (1/4″) thickness or greater are required to perform accurate shear tests. Shear test specimens are further discussed by Renton [35–37].

In an AFML Report [36], Renton utilizes the Buckingham Pi theorem to develop additional parameters proportional to the adhesive shear and normal stresses: $A_{11}\eta/C_aL^2$ and $D_{11}\eta/E_aL^4$, respectively, where L is the length of the adhesive bond in the single lap joint. Also, a poll of fifteen researchers clearly points out the agreement with and the advantages of the thick adherend shear test piece compared with alternatives.

In 1979, Jonath [41] described the Iosipescu specimen, Figure 8.12, used to measure adhesive shear strength. Flaggs has analyzed the specimen by finite element methods, and the adhesive is almost in pure constant shear. However, to date displacements have not been measured; hence, it has not been possible to measure the elastic modulus, the proportional limit or the strain to failure. Incidentally, much earlier, Iosipescu [49] used a specimen similar to Figure 8.12 to measure the shear strength of materials other than adhesives. This method is still used for that purpose, including the Army Materials and Mechanics Research Center, who is using it to measure the shear strength of graphite epoxy composite materials. Kutscha and Hofer [6] recognized the potential of the Iosipescu specimen for measuring the shear strength of joints and the shear strength of adhesives. They note the difficulty in fabrication but

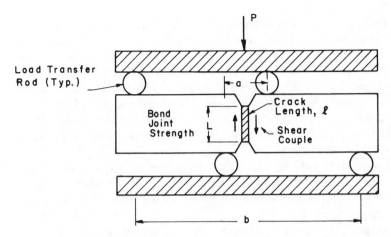

Fig. 8.12. Iosipescu Test Specimen

remark that the difficulty is less than fabricating tubular torsion specimens. Also they note that with proper instrumentation the Iosipescu specimen could be used to measure the shear modulus.

There is no generally accepted adhesive tensile property specimen or test procedure. However, Renton and Vinson [11,13] developed a test piece and procedure, which provides tensile strength and modulus data. Further research was conducted by Renton [36] in 1978 and he states that further development of a satisfactory adhesive tensile test is very much needed.

Adhesive material property data have been published in several papers and reports such as Lehman and Hawley [7], Grimes [10], and Renton and Vinson [11,13,38]. More recently, the PABST Program [39] investigated several adhesive systems and eventually chose the American Cyanamid FM73M as the adhesive to use in this large program in bonded aircraft structures because it exceeded all of the military required environmental requirements. Recently, Flaggs and Vinson [40] provided shear property data for both room temperature and 212°F for nineteen adhesive systems, at negligible moisture level. In all of the work of Renton, Flaggs and Vinson the thick adherend test piece was used, with the ratios discussed above.

Most recently, Flaggs and Crossman [28] published viscoelastic properties of FM73M adhesive, and analyzed viscoelastic effects on stresses in single lap joints. The development of such methods of analysis is very new. As Renton [36] stated in September 1978, in his review of 198 documents, viscoelastic analysis per se is absent in all the articles reviewed. However, as early as 1965 Miller, Jergens and Plunkett [50] made some measurements of the complex shear modulus of adhesives using the forced vibration technique. He stated that it will be necessary to characterize the viscoelastic properties of adhesive materials, especially when they are used in a thermal environment. However, their work apparently did not attain wide distribution. Oplinger [25] in 1975 stated that the need for investigating the adhesive viscoelastic properties in order to identify approaches to joint fabrication which will minimize thermal stresses at typical temperatures for which the structure is designed is obvious. Also, in 1977 Sen [33] assumed a viscoelastic adhesive in his double lap joint study.

Recently, Jurf [99,100] experimentally determined viscoelastic shear properties for two epoxy adhesives, FM73M and FM300, over a range of temperatures and hygrothermal environments using the thick adherend test piece of Figure 8.11.

Stenersen [42] recently published elastic material property data on various polyimide-adhesives at both 25°F and 600°F.

Oplinger [25] states that the application of a fracture mechanics approach as a means of predicting crack growth rates in the fatigue of bonded joints appears desirable. Williams [61] has discussed fracture mechanics approaches appropriate to adhesive joints. An alternative approach that of DeVries [62] can also be used.

Also of interest is the fact that electrically conducting silver filled adhesives have been used successfully to provide grounding between satellite components [43]. Also thermally conductive, non electrically conductive aluminum oxide filled adhesives have been used to install temperature transducers on satellites. At present there are no commercially available electrically conductive, non thermally conductive adhesives, but they could be used for a multitude of applications when and if they are developed.

The surface has just been scratched in providing elastic and viscoelastic properties of important adhesive systems at various temperatures and moisture levels. Also, considerable research is needed to develop better shear and normal displacement transducers to obtain accurate adhesive tensile and shear stress strain curves.

8.2.2.8. General

Thrall [39] stated in 1979 that there are essentially no primary aircraft structures of American design in commercial service today which rely on an adhesive to carry or transfer 100% of the local structural loads, although many so called secondary structures utilize adhesive bonding, but even there in many cases mechanical fasteners also installed through the joints (in the vernacular these are called chicken rivets or chicken bolts). In Europe however several primary structures have been joined through adhesive bonding: the Fokker F-27 and F-28 (1955), the de-Havilland Comet (1949) and the Hawker-Siddely Trident IIIB.

In the PABST full scale structural test, after 68,000 pressure cycles on a riveted structure, seventy five cracks were identified; after the same number of cycles on the bonded structure, only seven fatigue cracks developed, and each of them was associated with a fastener hole.

Portions of the above review were presented as a paper at the ASME 100th Anniversary Centennial Conference [63].

8.2.3. General Analytical Approach to Study Adhesively Bonded Joints.

Consider two or three flat panels joined together through adhesively bonded joints, subjected to uniform in-plane mechanical loads in the x direction, and with a continuous temperature distribution and a continuous moisture distribution varying in the x direction and in the thickness direction, z. If these panels are sufficiently wide in the y direction then the combined structure may be considered to be in a state of plane strain in the y direction.

For such cases, many different joints may be analyzed by developing a general solution for the portion of the composite material adherend depicted by the Wetherhold-Vinson model in Figure 8.4. The laminated element shown is subjected to stress resultants N_1 and N_2, stress couples M_1 and M_2, and shear resultants Q_1 and Q_2. The adherend elements are also subjected to a given distributed normal load on the upper and lower surfaces $p_U(x)$ and $p_l(x)$, distributed shear loads on the upper and lower surface $\tau_U(x)$ and $\tau_l(x)$, and a continuous temperature distribution $T(x, z)$ and a continuous moisture distribution, $m(x, z)$. While solutions to the Wetherhold-Vinson model are general for most practical adherend stacking sequences, they are restricted to those of *midplane* (*x-y* plane) *symmetry* which are in a state of plane strain in the y direction. Thus, this solution is an analytical finite element for the adherends in any of the joint configurations shown in Figures 8.1, 8.5, 8.6, 8.7 and 8.9.

The adhesive in these configurations is modelled using an elastic film approximation used previously by Renton and Vinson [11,13,18] as well as several other researchers, and modified herein to include hygrothermal effects. These effects are important to polymer matrices in composite adherends, and must also be considered important for the pure polymer adhesives. By combining the analyses of the adherends and the adhesive, the joint configurations of Figures 8.1, 8.5, 8.6, 8.7, and 8.9 may be analyzed.

Wetherhold-Vinson Analysis for an Adherend Element

From Reference 23, the governing equations for a portion of an adherend subjected to all the loads discussed above are given below, wherein each symbol could be subscripted i, in order to use it subsequently as a building block in analyzing each of the various joint configurations.

$$\frac{dN_x}{dx} + \tau_u - \tau_L = 0 \tag{8.4a}$$

$$\frac{dQ_x}{dx} + \sigma_u - \sigma_L = 0 \tag{8.4b}$$

$$\frac{dM_x}{dx} - Q_x + \frac{b}{2}(\tau_u + \tau_L) = 0 \tag{8.4.c}$$

$$M_x = -D_{11}\frac{d^2w^0}{dx^2} + F\frac{d^3\phi}{dx^3} + G\frac{d\phi}{dx} + H\frac{d^3\tau_L}{dx^3} + I\frac{d\tau_L}{dx} + \bar{H}\frac{d^3\tau_u}{dx^3}$$

$$+ \bar{I}\frac{d\tau_u}{dx} + \bar{h}(x) - M^T(x) - M^m(x) \tag{8.4d}$$

$$N_x = A\frac{d^3\phi}{dx^3} + B\frac{d\phi}{dx} + C\frac{d^3\tau_L}{dx^3} + D\frac{d\tau_L}{dx} + \bar{C}\frac{d^3\tau_u}{dx^3} + \bar{D}\frac{d\tau_u}{dx} + E\frac{d^2\sigma_L}{dx^2}$$

$$+ \bar{A}\frac{du^0}{dx} - N^T - N^m + \bar{E}\sigma_L + h^*(x) \tag{8.5}$$

$$Q_x = K_5\phi(x) \tag{8.6}$$

In the above, N, M, and Q are the usual stress resultant, stress couple and shear resultant quantities as defined in numerous references (for example Reference 26) once they have been determined, a conventional laminate analysis may be used to determine the stresses in each ply. The quantities w^0 and u^0 are the midplane lateral and in-plane displacements respectively, and ϕ is proportional to the rotation of the midplane. Thus, there are six equations and six unknowns for this problem, which includes all of the loads shown in Figure 8.4. However, in what follows we will treat this problem as a building block where the loads are, in general, unknowns which must be determined. The lettered constant coefficients in 8.4 − 8.6, which involves material properties and geometry, are defined in Reference 23 and are too lengthy to be included herein.

Analysis of the Polymer Adhesive

Consider the adhesive layer in a single lap joint as shown in Figure 8.1 and 8.3. If we denote the displacements of the upper adherend and the lower adherend with subscripts 2 and 3 respectively, then the normal strains in the z direction and the shear strains in the $x - z$ plane can be written as

$$\frac{w_2(x_2, -h_2/2) - w_3(x_3, h_3/2)}{\eta} = \frac{\sigma_0(x)}{E_a} + \alpha_a\Delta T + \beta_a m \tag{8.7}$$

$$\frac{u_2\left(x_2, -h_{2/2}\right) - u_3\left(x_3, h_{3/2}\right)}{\eta} = \frac{\tau_0(x)}{G_a} \tag{8.8}$$

where σ_0 and τ_0 are the unknown normal and shear stresses in the adhesive respectively; E_a and G_a are the modulus of elasticity and the shear modulus of the adhesive in film form as determined, for example, by Renton, Flaggs and Vinson [19]; η is the thickness of the adhesive in the z direction; α_a and β_a are the coefficient of thermal expansion and hygroscopic expansion as discussed by Pipes, Chou and Vinson referenced earlier; ΔT is the change in temperature between the temperature of the material point considered and the "stress-free temperature," and m is the moisture content in weight percent existing in the adhesive.

Thus, it is seen that for each adherend element there are six governing differential equations, given by 8.4 through 8.6. Also there are two equations describing the elastic behavior of the adhesive in a mechanics of materials sense. These, then, comprise all of the equations necessary to study the behavior of any of the joint configurations described previously. In the instance of a single lap joint, consisting of four adherend elements and one bond line, there will be 26 equations and 26 unknowns. For a double lap joint, there are 6 adherend elements and two bond lines, hence 40 equations and 40 unknown functions. Even more equations and unknowns are required for other configurations.

8.2.4. Design Considerations

It is seen that the design and analysis of adhesively bonded joints is very complex. If an analytical approach is used it involves at least 26 equations and 26 unknowns, and after the roots of the equations are found, a computer program such as BOND 4 is needed to do design, analysis or optimization studies.

Computer programs listed previously are available. Each has different assumptions, and concerns different configurations. One of the problems involved is the nature of the shear stress distribution, see Figure 8.2, which often requires double precision on the computer.

For very simplified preliminary design studies, the excellent paper by Oplinger [22] can often be used, at least qualitatively.

Some general comments can be made.
1. Maximum shear stresses and normal stresses can be reduced by making the D_{11} and A_{11} of the adherends as high as possible.
2. Regardless of the materials used, the values of A_{11} and D_{11} of the adherends on each side of the bond should be equal. Otherwise there will not be symmetry of either the adhesive shear and normal stress about the mid-length ($x_2 = x_3 = \frac{1}{2}L$), with higher stress values resulting from the asymmetry.

3. As Oplinger [22] points out, there is a length of bond line beyond which no load capacity increase occurs, due to the nature of the shear stress distribution, Figure 8.2.

4. If the bond line length is made larger and larger, then failure or yielding will occur in the adherend.

5. Hart-Smith [65] points out that if there is a debond of length L_1 at the end of a bond line of length L, then approximately one can simply analyze the same adhesive bond, but of length $L - L_1$. That paper also gives much other practical design information.

Many other factors must be studied more in detail such as the effects of viscoelasticity and plasticity in adhesively bonded structures. As stated previously, Flaggs and Crossman [28] were the first to study the effects of viscoelasticity. Recently Delale and Erdogan [66] have also contributed to this area.

Many of the Hart-Smith references listed herein involve the study of plasticity effects on adhesive stresses. However, his work to date deals with elastic-perfectly plastic characterization of the adhesive shear stress strain curve, with the normal adhesive stresses remaining elastic. Also, he does not permit the shear stresses to go to zero at the end of the bond line, see Figure 8.2, because the plasticity model used does not permit that.

8.3. Mechanical Fastening

8.3.1. Introduction

Although, almost all engineers philosophically agree that adhesive bonding is superior to mechanical fastening, primarily because of continuous joining, no fiber cutting, and no stress concentration, nevertheless for structural components which must be removable or easily replaceable mechanical fastening has its very important place.

8.3.2. Review of Mechanical Fastening Technology.

The early research on mechanical fasteners effects dealt with joining isotropic metallic or plastic structural components. In 1928, Bickley [67], in an excellent paper, studied the state of stress around a circular hole in an infinite elastic plate of uniform thickness subjected to prescribed in-plane tractions acting on the periphery of the hole. He obtained general solutions, and then concentrated on cases wherein the specified tractions might occur in practice. One case studied was prescribing a normal pressure varying as cos θ over the surface of the hole, and he states that will provide the stress distribution around a hole into which a rivet has been driven, when the rivet has been pulled sideways by a force in the plane of the plate. As will be discussed later, this is no longer considered to be an accurate assumption. Tables and figures of the stress distributions are given by Bickley. He also provides complete solutions for the most general in-plane tractions around the hole, therefore, there are solutions available for a more carefully modeled characterization of a mechanical fastener. He also provides solutions for a single force F, acting on the hole at $\theta = 0$. This solution can be used to form a Green's function, useful in prescribing any radial traction on the boundary. He also studied the effects of a uniform radial pressure along an arc, and a traction varying as $\cos \theta$ over half of the boundary, and provides explicit solutions and graphs of all stress components. He also studied the stress distribution approximating a rivet which is too small for the hole, and a tangential traction involving $\sin^3\theta \cos \theta$ because he states there will be some tangential friction between the rivet and the hole. Lastly, he compares his analytical prediction with experimental results, and the agreement is quite good generally.

Howland [68], in 1930, studied the stress distribution in a finite strip with two parallel straight boundaries with a central rivet hole. He shows that the maximum value of the circumferential stress at $\theta = 0$ is 5.5 times the tensile stress at infinity when a rivet is pulled with a force P in the $\theta = 90°$ direction, when the rivet hole occupies half the width of the strip. Tables and figures are provided for all solutions. [For basis of comparison, if the strip is subject to uniform tension in the $\theta = 90°$ direction, and the rivet hole is traction free, the maximum values of the circumferential stress occur at $\theta = 0°$ and are three times the uniform tensile stress for a small hole but rises to 4.3 when the hole occupies half the width of the strip]. Howland also states that when there are rivets on either side of the one considered, it is likely that the maximum stress above will be reduced by more than one third. Howland also discusses earlier work by Filon and Jeffery which all not be reviewed herein.

In 1935, Knight [69] studied the effects on stresses of neighboring rivets and boundaries of the isotropic plate. He also studied the cases wherein the bolt load is parallel to the edge of the strip and when it is perpendicular to the edge. Figures and tables of results are given. He also points out that some of Howland's [68] figures require slight corrections.

Twenty years later Theocaris [70] presented an exact solution for the stress distribution resulting from loading a perforated strip in tension, in which the hole is filled with a rigid pin, a paper based on his doctoral dissertation. The problem is similar to the problem studied by Howland, and although Theocaris references other work of Howland, he does not reference [68]. Theocaris provides his results in the form of stress concentration factors which he provides in graphical and tabular form.

He found that the maximum principal stress exists at 90° from the pin loading direction at the rim of the hole, and is 6.5 times the average stress when the hole radius is 10% of the strip width, and becomes a minimum of about 5 times the mean stress when the hole radius is 20% of the plate width. This differs from the results of Howland [68]. Also, the influence of the hole on stress distribution extends only in a region of diameter equal to 1.5 times the strip width. He also found that at the strip edge at location of minimum cross-section the tensile stress is about one-half of the average stress, regardless of the ratio of hole radius to strip width.

In all of the above early studies only isotropic, thin classical flat structures were studied under static loads. Neither dynamic or thermal loads were considered, nor were nonlinear effects. They are included to provide an historic background.

In June 1969, Kutscha and Hofer [6], of IITRI presented the first comprehensive treatment of mechanical fasteners in composite structures. Their analytical and experimental study included experimentally studying the load distribution and joint strength in fiberglass bolted joints and comparing them with analytical solutions. They also provide an inclusive literature survey for design procedures for joining with mechanical fasteners. They found very few references dealing with joining fiberglass and note them on page 372 of their report.

Kutscha and Hofer conclude from their AFML sponsored study that in 1969 the most effective way to deal with bolted joints in *isotropic* materials is to employ semi-empirical methods for design. For *composites* the failure modes are so different (tensile, compressive, shear buckling or mixed in any combination) a complete stress analysis and failure envelope (in four dimensions) is required to even develop the semi-empirical procedure to design a given class of joints. They also list six recommendations for further research which are inclusive.

A short time later, June 1969, Lehman, Hawley and their colleagues at McDonnell-Douglas published an inclusive technical report [7], presenting the results of their studies under AFFDL sponsorship on bonded and bolted joints in composite panels under static and dynamic loads. In their extensive program they first determined the properties of boron and S-994 fiberglass laminates using Narmco 5505 resin in tension, compression, in-plane shear, interlaminar shear, and pin-bearing properties. These material properties are presented in the report as well as a description of the test methods used. The bolted joint concepts. they studied included laminates with plain holes and with reinforced holes using either a composite or a steel shim reinforcement, or steel bushings. They found the factors affecting bolted joint failures are laminate strength and stiffness, edge distance, fastener pitch, laminate thickness and fastener diameter. One interesting result from their study is that a combination bolted-bonded joint performed better than bolted joints or bonded joints alone, primarily because the combination causes a fundamental change in failure mode. They found that bolted joints in composites are lighter than bolted joints in aluminum. However, bonded joints were found to be superior to bolted joints on a weight-efficiency basis.

They also concluded that their linear discrete element analysis did not adequately predict load-deformation characteristics of bolted joints. Agreement between test and linear theory for the basic fastener-sheet combination was poor. The discrepancy is probably due to nonlinear behavior due to local filament buckling and matrix plasticity caused by high bearing stresses. They conclude that an extremely sophisticated nonlinear analysis is required to characterize the load-deformation behavior of bolted joints.

The bolted joints studied by Lehman and co-workers included both single and double lap configurations. The single lap joints were not penalized greatly strengthwise due to the asymmetry of the joint. In the double lap joint the use of bushings reduced the joint weight efficiency, so this is not a good idea. Both the insertion of metal shims and thickened end designs resulted in weight effective joints. Although more difficult to fabricate, the shimmed joint produces a very compact, high-strength joint. The thickened-end configuration, however, is a good general purpose design. As stated above the addition of adhesive bonding to a bolted joint gave strengths that were greater than similar joints using either bonding or bolting only. Specifically, bolted joints without bonding were less than 1/3 as strong in the fiberglass laminates and less than 1/5 as strong in the boron laminates. Shearout failures were prevalent in all bolted joints except for the shim-reinforced and the combination bolted-bonded joints, even when the joint proportions were selected to produce bearing failures. The use of whiskers as a resin additive did not increase the strength of the joint.

Shear stresses developed at failure decrease as edge distance increases. Thus, increasing edge distance is an inefficient method of increasing joint strength. However, for a given edge distance, joint strength increases approximately linearly with composite thickness until bolt bending effects reverse this trend.

Edge distances of 4.5 times the fastener diameter were required to achieve a balance between shear-out and bearing strengths. Tensile failures through the fastener holes were precluded by a side distance of twice the fastener diameter. A thickness-to-diameter ratio of 0.8 results in maximum bearing strength. Maximum shear-out stresses were developed in a laminate containing $0°$ and $\pm 45°$ layers in which 2/3 of the laminae were $\pm 45°$ to the load axis.

In their bolted joint tests, Lehman indeed used bolts rather than pins so that the effects of bolt clamping friction were included. Consistent surface conditions and bolt torques were maintained to minimize scatter. Standard torques were used resulting in bolt stresses between 30 and 40 ksi. Standard specimen width was 1″, 3/16″ and 1/4″ bolts were used, hence s/D ratios are 2.63 and 2.00 respectively where S = distance from bolt center-line to the edge of the test piece. Lehman states that test results indicate that the full bearing stress will not be developed if D/t (where t is the laminate thickness) exceeds a value of two. Hence, he chose a normal laminate thickness of 0.120 inches. He also stated that published test results show that the bearing stress increased with e/D until a ratio of 5 is reached, where e = edge distance. He chose 1/2″, 3/4″ and 1-1/4″, such that the e/D ratio varied from 2.63 to 6.58. His tests show that the bolted joint strength increases linearly with e/D ratio up to a value of 4. Beyond that value, increasing the e/D ratio did not appreciably increase joint strength, regardless of t/D.

Lehman proposes the following empirical equation to determine bolted joint strength

$$\frac{P}{Dt} = K_n\left(\frac{e}{D}\right)[1 - e^{-1.6(s/D - 0.5)}][1 - e^{-3.2(t/D)}], \qquad 0 < \frac{e}{D} < 4 \quad (8.9)$$

where K_n is an experimentally determined coefficient in units of psi.

For higher edge distances

$$\frac{P}{Dt} = K_n F_{\text{BRU}} \qquad e/D > 4 \tag{8.10}$$

where K_n is a coefficient, and F_{BRU} is the ultimate bearing strength of the material. Both equations above are based on experiments for $s/D > 2$, and laminates of $0°$ and $\pm 45°$ only, wherein no tension failures occurred (only shear-out and bearing failures occurred). Lehman concludes like Kutsche and Hofer that at present (1969), semi-empirical methods are the most effective approach to rational joint design in composite materials.

Fatigue tests were conducted on both single and double lap bolted joints at a stress ratio of $+0.05$, wherein the maximum loads ranged from 60% to 80% of the static load capacities to reduce the possibility of runout. Boron reinforced specimens were cycled at 1800 cycles per minute and fiberglass-reinforced specimens were cycled at 900 rpm to minimize the temperature rise. Even so the single lap joints suffered a $50°F$ temperature rise, and the double lap joints stabilized at $86°F$. Lehman found that the bolted double lap specimens produced the longest fatigue life (over the bolted single lap, and single and double lap adhesive bonds, adhesive bonded scarf and adhesive bonded step lap joints) and in each case failure occurred in the aluminum alloy members through the bolt hole. The combination bolted and bonded joint of boron also performed well in fatigue.

In 1970, Harris, Ojalvo and Hooson [71] of Grumman Aerospace performed a comprehensive analytical study on bolted joints using plane stress finite elements for both the elastic and plastic regions. They analyzed the load-deflection behavior of single fastener joints, residual stress distribution in plates with squeeze rivets, the effect of fastener bending and shear deformation on the bearing stress distribution between the fastener and plate, and made predictions of fatigue life of typical mechanically fastened joints.

About the same time Gould and Mikic [72] of MIT also used finite element methods to compute pressure distributions, the contact zones between bolt and plate and the radii at which flat and smooth asymmetric linear elastic plates of various thicknesses will separate. They also measured the radii of separation by two methods. One utilized autoradiographic techniques and the other measured the polished area around the bolt hole caused by sliding. The results were that the finite element techniques produced results in agreement with the experimental data, and both yield smaller zones of contact than those indicated in the literature. The analysis logic and the computer programs and users manual are given in the reference.

Although the above two references do not deal with composite structures, they are included because the techniques used may be useful.

In 1971, Waszczak and Cruse [73] published a paper providing means through finite element methods to predict the elastic stress distribution around pinned joints in anisotropic structures. The effect of the pin was represented by a half cosine radial pressure distribution. Failure modes were predicted using a distortion energy principle and the assumption that damage to any ply represented laminate failure.

In the next year, Cruse and Waszczak, along with Konish [74] continued their research and investigated the effects of sharp cracks and loaded holes on the strength of composite material structures, using a simple model obtained from lamination theory. The strength of bolted joints in boron and graphite epoxy plates were evaluated. They also made predictions regarding fracture in advanced composites. Their results conclude that in order to employ linear elastic fracture mechanics, factors such as the lower bound on crack length and the material dependence on the finite dimension correction must be determined.

Also in 1972, Dickson, Hsu and McKinney [75] studied fatigue phenomena in both bonded and bolted joints in advanced composite materials, and provided a means for predicting the elastic stress distribution around pinned joints in composites.

In 1973, Waszczak and Cruse, continuing their research, produced an AFML technical report [76], which provides a general optimization procedure based on their previous research, utilizing laminate failure modes and minimum weight considerations. Again the half cosine radial pressure distribution assumption was made.

In May 1974, Oplinger and Gandhi [77] published analytical results for the elastic response and the failure of mechanically fastened orthotropic panels with either single or multiple fasteners. An interactive approach was used to determine the contact arc, a nonlinear aspect of the problem. The joint configurations studied included the single pin, a periodic array in finite rectangular plates, and an isolated pin in a infinite plate. The least squares collocation method with a complex variables formulation of the two dimensional problem was the approach used. They found that an optimum value of $S/D \approx 2$ corresponds to a minimum in the ratio of peak net section tensile stress to applied stress, where $2S$ is the distance between pin centerlines and D is the pin diameter.

In September 1974, Murphy and Lenoe [78] published a unique annotated bibliography on previous research on structural joints of all kinds, as well as interfaces. It was the first bibliography and hopefully the present document will be its update in the area of bolted joints only. This present work does not include references of [78] that do not deal with composites.

Also in September 1974 the Army Materials and Mechanics Research Center held a Symposium dealing primarily with the role of mechanics in the design of structural joints. At that meeting Van Siclen [79] of Northrup Aircraft spoke on his work in developing design procedures for bolted joints in graphite epoxy laminates. His research had a dual purpose: one, to develop joint allowables and two, to evaluate the merits of alternative reinforcing concepts for improving joint strength. The approach Van Siclen used, similar to Lehman earlier, was to obtain and evaluate actual joint test data for Thornel 300/PR286, and from the data to establish semi-empirical procedures for predicting joint strength as a function of all pertinent variables. The reinforcing concepts considered were metallic interleaves, externally bonded on metallic doublers, laminate cross-ply build ups and fiberglass "softening strips."

Van Siclen aptly states that "the main difficulty in designing mechanical joints in composite laminates is that the behavior of such joints in composites is not well understood. This is due in part, to the general complexities associated with composite materials and partly to the lack of understanding of mechanical joint behavior."

Van Siclen also lists several references which are cogent to bolted joints which are not discussed herein.

He discusses each of the geometric parameters which affect joint behavior which are edge distance (e), side distance (s), hole diameter (D) and laminate thickness (t), along with laminate properties which includes fiber orientation, stacking sequence, and the types of material systems.

He also points out that the type of fastener used has a great influence on joint strength. For example when a countersunk fastener is used rather than a protruded head fastener, the bearing strength of the composite joint may be severely reduced.

In the Phase I test program the various parameters that affect the behavior of mechanical joints in graphite-epoxy laminates were evaluated. Both single and double lap joints were used with one protruded head fastener. Specimen length was 5.0″, and specimen width was twice the side distance.

In Phase II tests, consisting of 54 double lap specimens, alternate joint reinforcing techniques were studied. Details of the reinforcements are given in the paper.

Concerning net tension failure, the allowable stress for a laminate is significantly reduced as the s/D is increased. So an s/D greater than 3 or 4 will not significantly increase the net tensile load a joint can carry.

Following Lehman [7], Van Siclen empirically determined the net tension stress concentration factor, K_{tc}, to be

$$K_{tc} = 1 + A\left[\left(\frac{D}{2s}\right)^{-0.55} - 1\right]\left(\frac{e}{2s}\right)^{-0.5} \tag{8.11}$$

where A is an experimentally determined constant for a particular laminate orientation. So, knowing the ultimate tensile strength of the unflawed laminate, F_x^{tu}, the value of ultimate tensile strength of the laminate with a bolted connection, F_{nt}, is given by

$$F_{nt} = \frac{F_x^{tu}}{K_{tc}} \tag{8.12}$$

Therefore the net tensile load a given joint can carry, P_{nt}, is given by

$$P_{nt} = F_{nt}(2s - D)t \tag{8.13}$$

where $2s$ is the specimen width.

As opposed to Lehman, Van Siclen found that the allowable shear-out strength for a $(0°, \pm45°, 90°)$ graphite-epoxy laminate is significantly reduced as the (e/D) ratio is increased.

For $(e/D) > 4$, bearing at the hole becomes the mode of failure. He also states that this data differs significantly with the data in the Air Force Design Guide which implies that shear-out strength is a constant value for a particular laminate orientation. He warns that "the use of constant shear-out equations can lead to excessively conservative designs or worse totally unconservative designs depending upon the e/D value for the specimen from which the shear-out strength was determined."

Van Siclen equation for the shear-out strength, F_{so} is:

$$F_{so} = A_1\left(\frac{e}{D}\right) + A_2 \tag{8.14}$$

where A_1 and A_2 are constants determined for a specific laminate orientation.

Now, knowing F_{so}, the allowable load on a fastener at shear-out failure, P_{so}, is given by

$$P_{so} = F_{so}Dt\left(\frac{2s}{D} - 1\right) = F_{so}(2s - D)t \tag{8.15}$$

Van Siclen uses the obvious equation for the allowable bearing load on the joint P_{BR}:

$$P_{BR} = F_{BRu}Dt \tag{8.16}$$

where F_{BRu} is the ultimate bearing strength for a given laminate.

From the Phase II tests, Van Siclen found that of the various reinforcements studied the minimum weight reinforcing approach is to use $\pm 45°$ graphite-epoxy plies in the joint area, a result similar to that found by Lehman, Hawley, and co-workers [6]. However, for minimum panel thickness the use of internally bonded titanium interleaves was the best approach. When the joint geometry must be kept to a minimum the use of metal layers, bonded either internally or externally is best, because $e/D = 2$ can be used instead of $e/D \approx 3$ for all other reinforcements. Again this conclusion is analogous to that reached by Lehman, Hawley and co-workers [6]. As for fabrication and machining the all graphite/epoxy is the easiest, while the most difficult is the use of internally bonded titanium interleaves.

For reinforcements by graphite epoxy buildup, titanium doublers, titanium interleaves, and unidirectional E-glass, the ratio of initial failure to ultimate failure load was close to unity. Only for the 181 woven fiberglass cloth with a high (e/D) ratio was the ratio 0.060. Hence, for highly loaded graphite-epoxy joints, fiberglass reinforcement should be used only in those cases where the need for such material has been dictated by other requirements, such as softening strips or damage arrestment strips.

Lastly, Van Siclen points out that joints fabricated using the congenital hole technique (holes layed up during laminate fabrication using a mandrel which is later removed) appear to offer a feasible alternative to machined holes, but more experience is needed. He also urges testing to be done under a random fatigue loading.

At the same AMMRC Symposium in 1974, Gatewood and Gehring [86] spoke of an inelastic redundant-force method of analysis for determining the mode and load at failure for axially loaded, multi-fastener, multi-plate splice joints under non-uniform temperature distributions. Unfortunately, the paper dealt only with metallic structures. Gatewood and Gehring utilize a dimensionless Ramberg-Osgood equation to depict the plate element stress-strain relationship and the local fastener-plate load deformation characteristics, as determined from two-fastener single lap test specimens.

Gatewood and Gehring give three references, [87,88,89] dealing with joints under thermal loads, but they are restricted to metallic structures and elastic deformations. Their contribution in this paper is treating nonlinear inelastic effects.

Also at the AMMRC Symposium of 1974, Oplinger and Gandhi [90] presented a paper on their analytical research regarding mechanically fastened fiber-reinforced plates. They consider the fasteners as rigid circular inserts, and both single pin as well as periodic arrays were considered. For comparison they obtained analytical results with and without friction between pin and plates. Their approach was a two-di-mensional elastic analysis using complex variables as in [13] in conjunc-tion with a least squares collocation scheme for satisfying the boundary conditions. In fact, this paper is partially a review of [13]. They found that iteration approaches are required: (1) for the determination of the contact angle between fastener and plate; (2) in the case in which friction is considered for determining the portion of the fastener hole that is free of slip; and (3) in the case of an array of fasteners, the determination of the net lateral strain which makes the plate free of net lateral stress.

The circular boundary adjacent to the fastener is divided into a region of separation where there are no normal or tangential stresses and a contact region over which a radial displacement corresponding to a simple translation of the fastener is the required condition. Moreover the latter region is divided into a non-slip region and a slip region where a linear relationship between shear and contact pressure must be satisfied.

They also use Hoffman's modified distortional energy criterion [91] for layer-by-layer failure considerations. This has the advantage over the more normal distortional energy formulation used by Waszczak and Cruse [9] because it takes differences in tensile and compressive strength into account. For this case the criterion is given by

$$\frac{\sigma_L^2 - \sigma_L \sigma_T}{S_{Lc} S_{Lt}} + \frac{\sigma_T^2}{S_{Tc} S_{Lt}} + \frac{S_{Lc} - S_{Lt}}{S_{Lc} S_{Lt}} \sigma_L + \frac{S_{Tc} - S_{Tt}}{S_{Tc} S_{Tt}} \sigma_t + \frac{\tau_{LT}^2}{T_{LT}^2} = 1 \quad (8.17)$$

where

σ	= normal stress
τ	= shear stress
S	= Allowable normal stress

T = allowable shear stress
subscript L = fiber direction
subscript T = normal to fiber direction
subscript t = stress in tension
subscript c = stress in compression

Waszczak and Cruse [9] earlier found that assuming a half cosine pressure distribution between fastener and plate was adequate for the cases they studied. Oplinger and Gandhi find that the half cosine pressure distribution is adequate also if $(e/D) > 2$ and s/D is not too large.

Oplinger and Gandhi conclude that the optimum fastener spacing is equal to about twice the pin diameter, and that to develop full joint strength the minimum edge distance should be twice the pin diameter. This is because peak stresses become independent of (e/D) for $(e/D) > 2$. This differs from the findings of Lehman & Hawley [6] who found the above true for $e/D > 4$. Differences between the response of a single pin and multipin configurations do not appear to be significant in commonly encountered laminates such as $0_2^\circ \pm 45^\circ$ configurations. For unidirectional laminates however, fastener interactions are important.

They found that friction effects between fastener and plate cause a significant reduction in cleavage stresses. This accounts for the lack of observed cleavage failure in $0_2^\circ \pm 45^\circ$ laminates that are predicted by analysis neglecting friction. Experimental efforts are needed to try to establish what coefficients of friction actually exist.

They also conclude that joint strengthening schemes involving foil interlayers should use materials which will provide sufficient shear stiffness to effectively load the foil interlayers in shear. Candidate materials that appear promising include boron vapor deposited on a polyimide substrate, and film forms of graphite.

Finally, Oplinger and Gandhi conclude that their methods of analysis need to be improved by (1) allowing for a nonlinear response (discussed earlier in Reference 26), (2) accounting for interlaminar shear and normal stresses as well as other three-dimensional effects, (3) restraining effect of cover plates, and (4) layer buckling. They observed that where bearing failures occur, they appear to be related significantly to three dimensional phenomena such as splitting due to deformation in the thickness direction. Also, interlaminar failures, which cannot be studied with their methods are common in $\pm 45^\circ$ laminates.

One year later, Oplinger [92] reported again on his research at an AMMRC Conference. He reconfirmed that he still feels adhesive joints are potentially more desirable than bolted or riveted joints, but that mechanical fasteners are required because of component disassembly.

He emphasizes, in 1975, that while prediction of stresses in fastened joints can be readily performed, there are deficiencies in currently available failure rules. He states that much of the research that remains to be done involves improved concepts for describing failure. He emphasizes that three dimensional stress analyses and analyses which allow for nonlinear material behavior in the laminate are essential to improved joint design and reliability.

He points out that plastic deformation in many laminate configurations has a significant effect on stress distributions and failure, and that recent experiments he has performed using Moiré surface strain measurements on pin loaded $0°/90°$ fiberglass test pieces support this.

Oplinger now points out that because of the need to treat friction effects that using a displacement boundary condition at the hole is preferable to assuming the half cosine radial pressure distribution, although the latter is more convenient.

He concludes simply that maximum joint strength for a row of fasteners is achieved by making $(s/D) = 1$, where $2s/D$ is the fastener spacing in the span direction and $(e/D) = 2$. However, he states, where shear nonlinearity is present (e/D) considerably larger than two may be required.

Oplinger points out many physical considerations, as enumerated in this paragraph. It is essential to introduce anisotropic failure criteria along with the stress analysis to draw valid conclusions about structural performance. Development of three dimensional crack stress analysis is required to treat effects of layer interaction. Shear failure in an individual ply amounts to a softening, which is bridged by fibers of the adjacent plies. Therefore, nonlinear stress analyses are called for to give better predictions of failure.

Oplinger again concludes that the use of displacement boundary conditions at the interface between panel hole and fastener is really preferable over assumed pressure distributions such as the "half cosine."

At the same Conference in September 1975, Ojalvo [93] presented a survey of methods of analysis for mechanically fastened splice joints. He pointed out that accurate analysis methods require a solution for the single-fastener problem that use a displacement type boundary condition at the hole edge rather than an assumed radial pressure distribution, and the methods must allow a variation in contact angle of the bolt or pin on the plates, thus supporting the conclusions of Oplinger.

In January 1976, Stockdale and Matthews [94] demonstrated that ultimate bearing loads are increased 40% to 100% in glass reinforced laminates by the clamping pressure of a bolt.

In April 1976, a paper by Kim and Whitney [95] dealt with hygrothermal (combined high temperature and humidity) effects on bolted joints. They investigated the effect of 260°F and 1.5% moisture by weight on pin bearing strength using a Thornel 300/Narmco 5208 laminate. The orientations studied were $[0_2/\pm 45]_{2s}$, $[90_2/\pm 45]_{2s}$ and $[0/\pm 45/90]_{2s}$. They found that the pin bearing strength at 260°F and 1.5% moisture was 40% less than the strength at room temperature and dry for all of the above orientations. They further found that there was no interaction between temperature and moisture, they each individually influence the strength degradation.

In April 1977, Quinn and Matthews [96] investigated the effect of stacking sequence in glass reinforced plastics on pin bearing strength. They measured the pin bearing strength using specimens which had eight plies, and mid-plane symmetry. They used 0°, 90°, $\pm 45°$ and $-45°$ plies in eight different combinations. They concluded that placing the 90° layers at or next to the outer surfaces increases the bearing strength. Also the ultimate failure mode was dependent upon stacking sequence. The layup $[90/\pm 45/0]_s$ was found to have the highest bearing strength, being some 30% stronger than the weakest lay up which was $[0, 90, \pm 45]_s$. However, the strongest layup also demonstrated the least fail-safe failure mode.

Allred [97] at the same time published the results of his research on the behavior of Kevlar 49 fabric-epoxy laminates subjected to pin bearing loads. His study indicated that Kevlar reinforced composites would cause problems when they withstood bearing loads in bolted joints. This experimental investigation involved $[0/90/\pm 45]$, $[0/90]$ and $[\pm 45]$ laminates. Laminate thicknesses varied from 0.075" to 0.300", using pins with diameters ranging from 0.125" to 0.500". Specifically, the laminates were made from DuPont Style 181 Kevlar 49 fabric (50 × 50 yarns/inch 380 denier, 8 harness satin weave) preimpregnated with U.S. Polymeric E-781 epoxy resin. Sample width was 1.50", and a distance of 0.75" was maintained between the hole and the specimen edge. The hole tolerance was 0.002".

Allred states that it is desirable to obtain not only the ultimate bearing strength but also the entire load (stress) – deflection curve, because the material yield strength from such a curve provides a better design allowable than the ultimate bearing strength often used. This is particularly true for structures subjected to repeated loading.

For these Kevlar specimens all samples underwent classical bearing failures with the outer layers on each surface delaminating, splitting and being pushed outward. No shear-out or net section tensile failures were observed until considerable bearing deformations had occurred.

Allred notes that load-deflection curves from the test all resembled each other and consist of five distinct regions, which he describes in detail, for pin loading in which lateral deformation of the composite is permitted. If the laminate is constrained then a load-deformation curve of the familiar form will occur.

Allred concludes that yielding in bearing occurs substantially below the ultimate bearing strength, and that the yield strength in bearing is very sensitive to stacking sequence (as also reported by Kim & Whitney). The quasi-isotropic configurations $[0, 90, \pm 45]_s$ exhibited the highest yield strength of those tested. Ultimate bearing strength, however, is not sensitive to stacking sequence. Both ultimate and yield strength in bearing exhibit an increasing stress concentration effect with increasing Pin diameter. Perhaps, most importantly, bearing strengths of the Kevlar 49/epoxy laminates are substantially below those reported for other high performance composites, being 10–20 ksi for bearing yield strength values. Therefore, Allred recommends that local reinforcement is desirable or necessary when designing a Kevlar reinforced composite component.

In July 1977, DeJong [98] wrote on analytical research he performed on pin loaded holes in elastic orthotropic and isotropic plates. He assumed a rigid pin, no friction, normal stresses only in a half cosine series (with his orientation of $\theta = 0$ it was a half sine series).

De Jong states that "although due to their complexity the problems of static joint optimization and of determining the behavior under cyclic loading can only be solved by experiment, knowledge of the theoretical stress distributions around pin-holes in composites may be very useful in the analysis of the test results."

De Jong made calculations for a unidirectional composite, a $[90_4, \pm 45]_s$ and a quasi-isotropic carbon fibre reinforced plastic. He states that assuming the pin to be absolutely rigid is reasonable for every laminate except the unidirectional orientation. Further, only for the isotropic and a slightly anisotropic case does the cosine (sine) distribution appear to be a good approximation. He found that the tangential stresses do not always have their maximum at a location 90° from the portion of the hole in the load direction.

De Jong states also that he could have calculated the stresses in the individual plies of the laminates and determined ultimate laminate strength by applying some failure criterion. He did not do this because as he says three dimensional effects are predominant at the edge of the hole and limit the significance of using his two dimensional method calculations.

He also concludes that because of its low transverse strength and relatively high tangential stress concentration a unidirectional carbon fiber reinforced plastic is not a suitable material for use in mechanically fastened joints.

8.3.3. General Design Approach

It is clear that the complexities associated with joints involving mechanical fasteners in composite material structures preclude a rational analytical solution at this time, and any time in the near future. To have an analytical solution would require a large finite element code which takes into account three dimensional effects, friction between bolt and the hole, hygrothermal effects, nonlinear laminate response, bolt bending, and other factors. Even then practical considerations such as hole tolerances, and bolt torque actually applied would limit the usefulness of the results for design.

Therefore from a design standpoint is appears that the best approach to use is that of Van Siclen [79], wherein simple easy to use equations can be used, which necessitate certain tests to be performed.

For any given structural component, hopefully the material system, and the laminae orientation and number are decided on due to overall loading and environment for the component, rather than letting the joints control such decisions. Thus, material system, laminate orientation and thickness are predetermined before joint decisions are made or considered.

For that chosen laminate, tests described by Van Siclen should be performed to determine F_{ns}, the net tensile strength, F_{so}, the shear-out, and F_{BR} the bearing strength for that particular laminate. In Van Siclen's [79] excellent paper, he presented the results of tests he performed on a $[0°, \pm 45°, 90°]_s$ graphite epoxy laminate. The resulting curves are given in Figures 8.13, 8.14, and 8.15.

These results then are easily used with Equations 8.11 through 8.16 to design, analyze and optimize any joint. If the structural component is to be subjected to hygrothermal loads, then these tests should be carried out under those conditions, because Kim and Whitney [95] have shown the deleterious effects of a hygrothermal environment. If there is a normal pressure due to bolt torquing then the tests should be performed with the

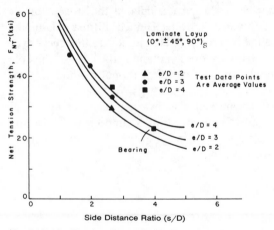

Fig. 8.13. Net Tension Strength as a Function of (s/D) and (e/D)

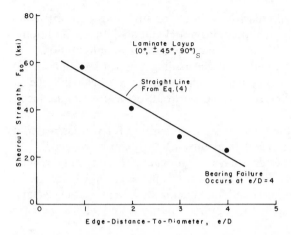

Fig. 8.14. Shearout Strength as a Function of Edge Distance

intended torques applied because Stockdale and Matthews [94] have shown that such pressure can increase bearing strength by 40% to 100%. Also, the actual bolts considered should be used in these tests because Van Siclen [79] points out the differences in strength when countersunk fasteners or protruded head fasteners are used.

With those strengths available one can then select a fastener size to transmit the required load. Keep in mind that it is generally agreed that $s/D = 1$ and $e/D = 2$ appear to be an optimum value in many cases. However using the Van Siclen equations one can, to a certain extent, optimize the mechanically fastened joint.

If the laminate is too thin to convey the required load then local

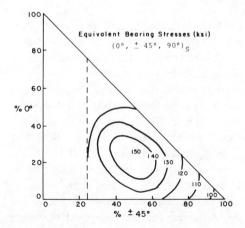

Fig. 8.15. Bearing Strength as a Function of Laminate Properties

278

stiffening is required. Alternative methods for stiffening were discussed by Lehman and Hawley [7] and Oplinger and Gandhi [90] and additional testing could be required in that case of the stiffened configuration.

If the structure is subjected to fatigue loading it appears a full cycle set of tests are required, although the preliminary design could be one in which there is no yielding in tension, shearout, or bearing at the mean fatigue load.

As a data bank on the three strengths is built up for various material systems and ply orientations the amount of testing can be reduced.

Since stacking sequence does have a significant effect on joint strength, if the stacking sequence for a structural component could be modified then the optimum design may call for a compromise between the stacking sequences best for the component primary purpose, and what is best for the joint strength.

Thus, the design of mechanically fastened joints is rather straightforward, and the optimization of joined structural components sufficiently complex that it is a challenge. In any case simple analyses can be used, but some testing is required.

It should also be remembered that Lehman and Hawley [7] found the addition of adhesive bonding to a bolted joint gave strengths that are greater than similar joints using either bonding or bolting only. Hence, in a design this also can be investigated using the Van Siclen approach discussed above.

It is interesting to note that none of the existing literature presented herein treats the problem of fasteners in composite material sandwich structures. However, again the Van Siclen approach can be used there also.

8.4. Problems

8.1. a. For the laminate discussed by Van Siclen and the results presented in Figures 8.13, 8.14 and 8.15, if each ply is 0.0055″ thick, using Figure 8.15 what total load (lbs) per bolt can be carried by a structure using 1/4″ diameter bolt? (i.e. $P_{BR} = ?$)

b. Using Figure 8.14, if the construction has an edge distance, e, of 3/4″, what side distance, s, is required to withstand the same load per bolt as in (a) above? (i.e. $P_{BR} = P_{so}$, $s = ?$)

8.2. a. For the laminate discussed by Van Siclen in Figures 8.13 and 8.14, namely $[0°, \pm45°, 90°]_s$, if each lamina is 0.0055″ thick, using Figure 8.15, what total load (lbs) per bolt can be carried by a structure using 5/8 diameter bolts? (that is $P_{BR} = ?$)

b. Using Figure (8.14), if the construction has an edge distance, e, of 3/4″, what side distance, S, is required to withstand the same load per bolt as in Problem 8.2 above? (that is, $P_{BR} = P_{so}$).

8.3. Consider a laminate composed of the composite of Figures 8.13, 8.14 and 8.15, $[0, \pm 45, 90]_s$, eight plys, hence $h = 0.044''$. If a $3/8''$ diameter bolt is used, with a side distance $s = 1''$, and $1''$ edge distance e, in which mode of the three will the panel fail due to the bolt load; what is that failure value?

8.4. Why is adhesive bonding potentially better for joining composite material components together than using mechanical fasteners?

8.5. List six different types of adhesively bonded joints.

8.6. What are three types of failure in mechanically fastened joints?

8.7. What makes a combination of mechanically fastened and a bonded joint superior to either a mechanically bonded joint or an adhesively bonded joint separately?

8.8. What are five good design rules for designing a good bonded joint in composite material structures?

8.9. What are five good rules to follow in designing good mechanically fastened joints in composites material structures?

References

1. Kuno, James K., "Structural Adhesives Continue to Gain Foothold in Aerospace and Industrial Use," Structural Adhesives and Bonding, *Proceedings of the Structural Adhesives Bonding Conference*, arranged by Technology Conferences Associates, El Segundo, California. (1979)

2. Szepe, F., "Strength of Adhesive-Bonded Lap Joints with Respect to Temperature and Fatigue" *Experimental Mechanics, Vol. 6* (1966): 280–286.

3. Volkersen, O. "Die Niet Kraft vertelung in Zug bean Spruchten Niet verb bind ungen mit Konstanten Lasch enguerschnitten," Luftfahrt forschungen, 15, (1938): 41–47.

4. De Bruyne, N.A., "The Strength of Glued Joints," *Aircraft Engineering, Vol. 16,* (1944): 115–118.

5. Goland, M., and E. Reissner, "Stresses in Cemented Joints," *ASME Journal of Applied Mechanics,* 11 (1947): A-17-A-27.

6. Kutscha, D. and K.E. Hofer, Jr., "Feasibility of Joining Advanced Composite Flight Vehicles, "*AFML-TR-68-391.* (January 1969).

7. Lehman, G.M. and A.V. Hawley, "Investigation of Joints in Advanced Fibrous Composites for Aircraft Structures," *AFFDL-TR-69-43, Vol. 1,* (June 1969).

8. Hart-Smith, L.J., "The Strength of Adhesive Bonded Single Lap Joints," *Douglas Aircraft Company IRAD Technical Report MDC-J0472,* (April 1970).

9. Hart-Smith, L.J., "Adhesive-Bonded Single Lap Joints," *NASA-CR-112236.* (January 1973).

10. Grimes, G.C., et al., "The Development of Nonlinear Analysis Methods for Bonded Joints in Advanced Filamentary Composite structures," *AFFDL-TR-72-97.* (AD 905 201). (September 1972).

11. Renton, W.J. and J.R. Vinson. "The Analysis and Design of Composite Material Bonded Joints Under Static and FAtigue Loadings, *AFOSR TR 73-1627,* (August 1973).

12. Renton, W.J. and J.R. Vinson. "Fatigue Response of Anisotropic Adherend Bonded Joints," *AMMRC MA-74-8.* (September 1974).

13. Renton, W.J. and J.R. Vinson. "The Analysis and Design of Anisotropic Bonded Joints, Report No. 2," *AFOSR TR-75-0125,* (August 1974).

14. Renton, W.J. and J.R. Vinson, "On the Behavior of Bonded Joints in Composite Materials Structures," *Journal of Engineering Fracture Mechanics, Vol. 7.* (1975): 41–60.

15. Renton, W.J. and J.R. Vinson. "Fatigue Behavior of Bonded Joints in Composite Materials Structures," *AIAA Journal of Aircraft, Vol. 12, No. 5.* (May 1975): 442–447.

16. Renton, W.J. and J.R. Vinson. "The Efficient Design of Adhesive Bonded Joints," *Journal of Adhesion, Vol. 7.* (1975): 175–193.

17. Renton, W.J., J. Pajerowski, and J.R. Vinson. "On Improvement in Structural Efficiency of Single Lap Bonded Joints," *Proceedings of the Fourth Army Materials Technology Conference-Advances in Joining Technology,* (September 1975).

18. Renton, W.J. and J.R. Vinson. "Analysis of Adhesively Bonded Joints Between Panels of Composite Materials," Journal of Applied Mechanics. (April 1977): 101–106.

19. Renton, W.J., D.L. Flaggs and J.R. Vinson. "The Analysis and Design of Composite Materials Bonded Joints, Report No. III." *AFOSR-TR-78-1512,* (1978).

20. Sharpe, W.N., Jr. and T.J. Muha, Jr. "Comparison of Theoretical and Experimental Shear Stress in the Adhesive Layer of a Lap Joint Model," *AMMRC MS 74-8.* (1974).

21. Srinivas, S., "Analysis of Bonded Joints," NASA TN D-7855, (N75-21673) (April 1975).

22. Oplinger, D.W., "Stress Analysis of Composite Joints," *Proceedings of the Fourth Army Materials Technology Conference-Advances in Joining Technology.* (September 1975).

23. Wetherhold, R.C. and J.R. Vinson. "An Analytical Model for Bonded Joint Analysis in Composite Structures Including Hygrothermal Effects," *AFOSR TR 78-1337.* (1978).

24. Liu, A.T., "Linear and Elasto-Plastic Stress Analysis for Adhesive Lap Joints," *University of Illinois T & AM Report 410,* (July 1976).

25. Allman, D.J., "A Theory for Elastic Stresses Adhesive Bonded Lap Joints," *Quarterly Journal of Mechanics and Applied Mathematics, Vol. XXX.* (1977): 4.

26. Humphreys, E.A. and Herakovich, C.T., "Nonlinear Analysis of Bonded Joints with Thermal Effects," *Virginia Polytechnic Institute and State University Report VPI-E-77-19,* (June 1977).

27. Vinson, J.R. and J.R. Zumsteg, "Analysis of Bonded Joints in Composite Materials Structures Including Hygrothermal Effects," *AIAA Paper 79-0798,* (1979).

28. Flaggs, D.L. and F.W. Crossman. "Viscoelastic Response of a Bonded Joint Due to Hygrothermal Exposure," *Modern Developments in Composite Materials and Structures,* ASME. (1979)

29. Dickson, J.N., T.M. Hsu, and J.M. McKinney, "Development of an Understanding of the Fatigue Phenomenon of Bonded and Bolted Joints in Advanced Filamentary Composite Materials, Vol. 1., Analysis Materials." *AFFDL-TR 72-64* (AD 750 132. (June 1972).

30. Hart-Smith, L.J., "The Strength of Adhesive Double-Lap Joints," *Douglas Aircarft Co. IRAD Technical Report MDC J036F.* (November 1969).

31. Hart-Smith, L.J., "Adhesive-Bonded Double Lap Joints," *NASA CR-112235,* (January 1973).

32. Keer, L.M. and K. Chantaramungborn. "Stress Analysis for a Double Lap Joint," *ASME Paper 75-APM-7,* (June 1975).

33. Sen, J.K., "Stress Analysis of Double-Lap Joints Bonded with a Viscoelastic Adhesive," *Ph.D. Dissertation, Southern Methodist University.* (May 1977).

34. Guess, T.R., R.E. Allred, and F.P. Gerstle, Jr., "Comparison of Lap Shear Test Specimens," *Journal of Testing and Evaluation, Vol. 5, No. 3.* (March 1977): 84–93.

35. Renton, W.J., "The Symmetric Lap Joints – What Good Is It? *Society of Experimental Stress Analysis Journal* (May 1976).

36. Renton, W.J., "Structural Properties of Adhesives." *AFML TR-78-127.* (September 1978).

37. Renton, W.J. and W.B. Jones, Jr., "Instrumented Test Methods for Adhesively Bonded Joints," *Proceedings of the 10th SAMPE Technical Conference.* (October 1978).
38. Renton, W.J. and J.R. Vinson. "Shear Property Measurements of Adhesives in Composite Material Bonded Joints," *ASTM 580.* (1975): 119–132.
39. Thrall, E.W., Jr., "An Overview of the PABST Program," *Structural Adhesives and Bonding, Proceedings of the Structural Adhesives and Bonding Conference arranged by Technology Conferences Associates,* (March 1979).
40. Flaggs, D.L. and J.R. Vinson. "Shear Properties of Promising Adhesives for Bonded Joints in Composite Material Structures," *AIAA Journal of Aircraft.* (January 1979).
41. Jonath, A.D., "Some Novel Approaches to the Study of Adhesive Interface Phenomena and Bond Strength," *Structural Adhesives and Bonding, Proceedings of the Structural Adhesives and Bonding Conference, arranged by Technology Conferences Associates,* (March 1979).
42. Stenersen, A.A., "Polyimide Adhesives for the Space Shuttle," *Structural Adhesives and Bonding, Proceedings of the Structural Adhesives and Bonding Conference, arranged by Technology Conference Associates.* (March 1979).
43. Reinhart, T.J., Jr., "Evolution of Structural Adhesives," *Structural Adhesives and Bonding, Proceedings of the Structural Adhesives and Bonding Conference arranged by Technology Conferences Associates,* (March 1979).
44. Hennecke, E.G. and Reifsnider, K.L., "Ultrasonic Testing Methods for Composite Materials," *Composites Technology Review,* Vol. 1, No. 2, Winter, 1979.
45. Green, A.T., "Acoustic Emission for NDT of Adhesive-Bonded Structures," *Structural Adhesives and Bonding, Proceedings of the Structural Adhesives and Bonding Conference arranged by Technology Conferences Associates,* (March 1979).
46. Kutscha, D., "Mechanics of Adhesive-Bonded Lap-Type Joints: Survey and Review," *AFML-TDR-64-298* (December 1964).
47. Kuenzi, E. and G.H. Stevens. "Determination of Mechanical Properties of Adhesives for Use in the Design of Bonded Joints," *U.S. Forest Service Research Note FPL-011,* (1963).
48. Rutherford, J.L., F.C. Bossler, and E.J. Hughes. "Capacitance Methods for Measuring Properties of Adhesives in Bonded Joints," *Rev. Scientific Inst., Vol. 39, No. 5.:* 666–671.
49. Iosipescu, N., "New Accurate Procedures for Single Shear Testing of Metals," *Journal of Materials, Vol. 2, No. 3* (September 1967).
50. Miller, H.E., J.L. Jergens, and R. Plunkett, "Measurements of Complex Shear Moduli of Thin Viscoelastic Layers," *University of Minnesota Institute of Technology Technical Report 65-4* (October 1965).
51. AVCO Corporation, "Evaluation of Test Techniques for Advanced Composite Materials," *Quarterly Report No. 2, Contract F33 (615)-67-6-1719, Air Force Materials Laboratory.* (January 1968).
52. Hartman, A. and P. De Rijn, "The Effect of the Rigidity of the Glue Line on the Fatigue Strength of 2024-T Aluminum Alloy Specimens with an Adhesive-Bonded Reinforcing Plate on Both Sides," *National Aero-and Astronautical Research Institute.*
53. Wang, D.Y., "The Effect of Stress Distribution on the Fatigue Behavior of Adhesive Bonded Joints," *ASD-TDR-63-93,* AFML. (July 1963).
54. Prokhorov, B.F., "Joints Used in Plastic and Composite Shipboard Superstructures, Deckhouses, Light Bulkheads, and Enclosures," Sudostroyenlye, #9, 1965, *U.S. Department of the Navy Translation No. 2076-b.:* 44–51.
55. Franzblau, M.C. and J.L. Rutherford, "Study of Micromechanical Properties of Adhesive Bonded Joints," *Second Quarterly Progress Report; Contract DAAA 21-67-C-0500,* (December 1967).
56. Ramberg, W. and W.R. Osgood, "Description of Stress-Strain Curves by Three Parameters," NACA TN 902 (July 1943).

57. Sharpe, W.N. and Muha, T.J., "Comparison of Theoretical and Experimental Shear Stress in the Adhesive Layer of a Lap Joint Model," *Proceedings of the Army Symposium of Solid Mechanics, AMMRC 74-9* (1974): 23–41.

58. Hart-Smith, L.J., "Adhesive-Bonded Scarf and Stepped-Lap Joints," *NASA Cr 112237.* (January 1973).

59. Erdogan, F. and M. Ratwani, "Stress Distribution in Bonded Joints," *Journal of Composite Materials, Vol. 5.* (July 1971): 378–393.

60. Barker, R.M. and F. Hatt. "Analysis of Bonded Joints in Vehicular Structures," *AIAA Journal, Vol. 11,* (1973): 1650–1654.

61. Williams, M.L., "Stress Singularities, Adhesion and Fracture," *Proceedings of the Fifth U.S. National Congress of Applied Mechanics.* (1966): 451–464.

62. DeVries, K.L., M.L. Williams, and M.D. Chang. "Adhesive Fracture of a Lap Shear Joint," *Experimental Mechanics* (March 1974): 89–97.

63. Vinson, J.R. "On the State of Technology in Adhesively Bonded Joints in Composite Material Structures." *Emerging Technologies in Aerospace Structures, Designs, Structural Dynamics and Materials,* ASME (1980).

64. Sawyer, J.W. and P.A. Cooper, "Analyses and Test of Bonded Single Lap Joints with Preformed Adhesives," *AIAA Paper 80-0773.* (1980).

65. Hart-Smith, L.J., "Further Development in the Design and Analysis of Adhesive Bonded Structural Joints." Douglas Paper 6922 presented at the ASTM SYmposium on Joining of Composite Materials. (1980).

66. Delale, F. and F. Erdogen, "Viscoelastic Analysis of Adhesively Bonded Joints." *NASA Contractor Report 159294.* (1980).

67. Bickley, W.G., "The Distribution of Stress Round a Circular Hole in a Plate," *Royal Society of London, Vol. 227A,* (July 2, 1928): 383–415.

68. Howland, R.C.J., "On Stresses in Flat Plates Containing Rivet Holes," *Proc. 3rd Int. Cong. of Applied Mech. Vol. II,* (1930): 74–79.

69. Knight, R.C., "The Action of a Rivet in a Plate of Finite Breadth," *Philosophical Magazine S. Vol. 19, No. 127,* (March 1935).

70. Theocaris, P., "The Stress Distribution in a Strip Loaded in Tension by Means of a Central Pin," *ASME Journal of Applied Mechanics,* (March 1956): 85–90.

71. Harris, H.G., Ojalvo, I.V. and R.E. Hooson, "Stress and Deflection Analysis of Mechanically Fastened Joints," *AFFDL-TR-70-49.* (AD 709221). (May 1970).

72. Gould, H.H. and Mikic, B.B., "Areas of Contact and Pressure Distribution in Bolted Joints," *NASA CR 102866.* (N70-41982) (June 1970).

73. Waszczak, J.P. and T.A. Cruse, "Failure Mode and Strength Prediction of Anisotropic Bolt Bearing Specimens," *J. Composite Materials, Vol. 5.* (1971): 421–25.

74. Cruse, T.A., Konish, H.J., Jr. and Waszczak, J.P., "Strength Analysis for Design with Composite Materials Using Metals Technology," *Proceedings of the Colloquium on Structural Reliability, the Impact of Advanced Materials on Engineering Design,* edited by J.L. Swedlow, T.A. Cruse and J.C. Halpin, Carnegie Mellon University. (1972): 222–236.

75. Dickson, J.M., T.M. Hsu, and J. McKinney. "Development of an Understanding of the Fatigue Phenomena in Bonded and Bolted Joints in Advanced Filamentary Composite Materials, Vol. 1," *AFFDL-TR-72-64,* (June 1972).

76. Waszczak, J.P. and T.A. Cruse. "A Synthesis Procedure for Mechanically Fastened Joints in advanced Composite Materials," *AFML-TR-73-145, Vol. II,* (AD 771 795). (September 1973).

77. Oplinger, D.W. and K.R. Gandhi, "Stresses in Mechanically Fastened Orthotropic Laminates," *Proceedings of 2nd Conference on Fibers, Composites and Flight Vehicle Design, AFFDL-TR-74-103.* (Sept. 1974): 811–842.

78. Murphy, M.M. and E.M. Lenoe, "Stress Analysis of Structural Joints and Interfaces-A

Selective Annotated Bibliography," *AMMRC Manuscript AMMRC MS 74-10.* (Sept. 1974).

79. Van Siclen, Robert C., "Evaluation of Bolted Joints in Graphite/Epoxy," *Proceedings of the Army Symposium on Solid Mechanics, September 1974,* (AD 786543): 120–138.

80. "Exploratory Application of Filament-Wound Reinforced Plastics for Aircraft Landing Gear," *AFML-TR-66-309.* (December 1966).

81. Nelson, W.D. et al., "Composite Conceptual Wing Design," *AFML-TR-73-57,* (March 1973).

82. Padawer, G.E., "The Strength of Bolted Connections in Graphite/Epoxy Composites Reinforced by Colaminated Boron Film," *Composite Materials: Testing and Design, ASTM STD 497.*

83. Padawer, G.E., "Film Reinforced Multifastened Mechanical Joints in Fibrous Composites," *AIAA Journal of Aircraft, Vol. 10, No. 9,* (September 1973).

84. Cherry, F.D., "The Elimination of Fastener Hole Stress Concentrations Through the Use of Softening Strips," *AFFDL-TR-72-130,* (September 1972).

85. *Advanced Composites Design Guide, Vol. II Analysis, Third Edition,* (January 1973).

86. Gatewood, B.E. and R.W. Gehring, "Inelastic Mechanical Joint Analysis Method with Temperature and Mixed Materials," Proceedings of the Army Symposium on Solid Mechanics, (AD 786-543). (September 1974): 193–210.

87. Gatewood, B.E., "Thermal Loads on Joints," *Thermal Stresses,* McGraw-Hill Book Company, New York (1957).

88. Switzky, H., M.J. Forray, and M. Newman, "Thermo-Structural Analysis Manual," WADD TR 60-517, (September 1960).

89. Lobbett, J.W. and E.A. Rob, "Thermo-Mechanical Analysis of Structural Joint Study," WADD TR 61-151, (January 1962).

90. Oplinger, D.W. and K.R. Gandhi, "Analysis Studies of Structural Performance in Mechanically Fastened Fiber-Reinforced Plates," Proceedings of the Army Symposium on Solid Mechanics, (AD 786-543). (September 1974).

91. Hoffman, O., "The Brittle Strength of Orthotropic Materials," Journal of Composite Materials, Vol. 1, (1967): 200–206.

92. Oplinger, D.W., "Stress Analysis of Composite Joints," *Fourth Army Materials Technology Conference – Advances in Joining Technology.* (1975).

93. Ojalvo, I., "Survey of Mechanically Fastened Splice Joint Analyses," *Fourth Army Materials Technology Conference,* (1975).

94. Stockdale, J.H. and F.L. Matthews, "The Effect of Clamping Pressure on Bolt Bearing Loads in Glass Fibre-Reinforced Plastics," *Composites,* (January 1976): 34–38.

95. Kim, R.Y. and J.M. Whitney, "Effect of Temperature and Moisture on Pin Bearing Strength of Composite Laminates," Journal of Composite Materials. (April 1976): 149–155.

96. Quinn, W.J. and F.L. Matthews, "The Effect of Stacking Sequence on the Pin-Bearing Strength of Glass Fiber Reinforced Plastic," J. Composite Materials, Vol. 11, (April 1977): 139–145.

97. Allred, R.E., "Behavior of Kevlar 49 Fabric/Epoxy Laminates Subjected to Pin Bearing Loads," *Sandia Laboratories Report SAND 77-0347,* (April 1977).

98. De Jong, Theo., "Stresses Around Pin Loaded Holes in Elastically Orthotropic or Isotropic Plates," *J. Composite Materials, Vol. 11,* (July 1977): 313–331.

99. Jurf, R.A., "Effects of Moisture on the Static and Viscoelastic Shear Properties of Epoxy Adhesives", MMAE Thesis, University of Delaware, 1984.

100. Jurf, R.A. and J.R. Vinson, "Effects of Moisture on the Static and Viscoelastic Properties of Epoxy Adhesives", *Journal of Materials Science,* October, 1984.GO-

Appendix 1. Micromechanics

A. Elastic properties

For orthotropic materials in a plane state of stress, four technical engineering constants are generally needed as input into the constitutive equations in order to model the material behavior. In order to develop the analytical tools capable of modelling these material response parameters the subject of micromechanics is introduced. This topic is based principally upon the determination of overall material properties of lamina in terms of their respective constituent properties and volume fractions.

Assumptions

Fiber
– Homogeneous *
– Isotropic
– Linearly Elastic
– Regularly Spaced *
– Aligned

Matrix
– Homogeneous
– Isotropic
– Linearly Elastic *

Lamina
– Homogeneous
– Orthotropic
– Linearly Elastic *

* Note: Most questionable assumptions

A number of approaches are available for calculation purposes to establish the material properties of composites. Among these are,

Analytical approaches

Mechanics of Materials Approach
Theory of Elasticity
– Self-Consistent Model
– Variational Principles
– Exact Solutions
Empirical Equations

Discussion of these different approaches can be found in a number of references as for example the following:

References

Mechanics of Materials Approach [1,2]
Theory of Elasticity
– Self-Consistent Model [3,4,5,6,7,8]
– Variational Methods [9,10,11,12]
– Exact Solutions [13,14,15,16,17,18,19]
Empirical Equations [20,21,22,23]

From the above approaches, two will be cited for further discussion as useful to students and design engineers.

Exact approach

The first approach employs so-called exact data. This may be properly labelled as a best method when compared with other available techniques. Inherent in this technique is the assumption that the fibers in the matrix are formally arranged in some fashion as for example the two patterns shown in Fig. 1. a,b. Other stacking/array sequences are also important and have been considered by investigators. An essential feature of this analytical methodology is the introduction of the so-called Representative Volume Element. The RVE represents the smallest reference control volume of material which contains a fiber (or fibers) surrounded by the matrix material. The RVE is considered to be both uniform and repetitive over the entire domain of the composite lamina with response characteristics representative of the composite. As indi-

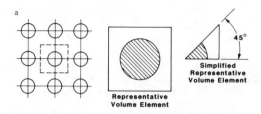

SQUARE ARRAY AND REPRESENTATIVE VOLUME ELEMENTS

Figure 1. a. Square Array And Representative Volume Elements

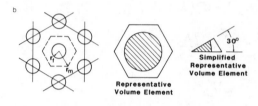

HEXAGONAL ARRAY AND REPRESENTATIVE VOLUME ELEMENTS

Figure 1. b. Hexagonal Array And Representative Volume Elements

cated in the assumptions for predicting the micromechanical behavior of composites, the effect of fiber arrangement/packing represents a fundamental building block in the modelling process. It appears that based upon investigation of this problem, that a hexagonal array model is more physically realistic than other geometrical arrayed models. The analytical process involved in thus determining the material properties follows the following process.

Exact analytical approach

Select an appropriate Fiber Array
Establish a representative RVE
Formulate an approximate Elastic Boundary Value Problem
Solve the formulated problem by means of
– Classical Stress Functions
– Series Solutions
– Numerical Methods (finite-differences)
Average the Elastic Fields to obtain the elastic constants
Results for calculating the four elastic constants using the "Exact" approach are noted below.

$$E_L = E_f - (E_f - E_m)V_m$$

$$E_T = 2[1 - \nu_f + \nu_m)V_m]$$

$$\times \left[(1 - C)\frac{K_f(2K_m + G_m) - G_m(K_f - K_m)V_m}{(2K_m + G_m) + 2(K_f - K_m)V_m} \right.$$

$$\left. + C\frac{K_f(2K_m + G_f) - G_f - K_m)V_m}{(2K_m + G_f) - 2(K_m - K_f)V_m} \right]$$

$$\nu_{LT} = (1 - C)\frac{K_f\nu_f(2K_m + G_m)V_f + K_m\nu_m(2K_f + G_m)V_m}{K_f(2K_m + G_m) - G_m(K_f - K_m)V_m}$$

$$+ C\frac{K_m\nu_m(2K_f + G_f)V_m + K_f\nu_f(2K_m + G_f)V_f}{K_f(2K_m + G_f) + G_f(K_m - K_f)V_m}$$

$$G_{LT} = (1 - C)G_m\frac{2G_f - (G_v - G_m)V_m}{2G_m + (G_f - G_m)V_m}$$

$$+ CG_f\frac{(G_f + G_m) - (G_f - G_m)V_m}{(G_f + G_m) + (G_f - G_m)V_m}$$

Nomenclature

ν_f, ν_m = Poisson's ratio of fiber and matrix respectively
K_f, K_m = Bulk Moduli for fiber and matrix respectively
G_f, G_m = Shear moduli for fiber and matrix respectively

$$K_f = \frac{E_f}{2(1 - \nu_f)}, \quad K_m = \frac{E_m}{2(1 - \nu_m)}$$

$$G_f = \frac{E_f}{2(1 + \nu_f)}, \quad G_m = \frac{E_m}{2(1 + \nu_m)}$$

The quantity C relates to fiber spacing in the matrix with $C = 0$ representing a single isolated fiber while $C = 1$ corresponds to full contact by the fibers in the matrix. A summary of curves for different values of C and fiber/matrix types and fiber volume fractions have been plotted in Figures 2-10. These plots are displayed for the three technical engineering constants which are functionally related to C, that is, E_T, G_{LT}, and ν_{LT}. It should be noted that from practical considerations that C is generally small in magnitude.

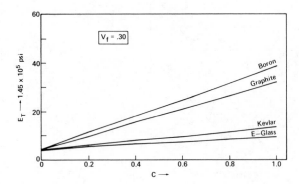

Figure 2.

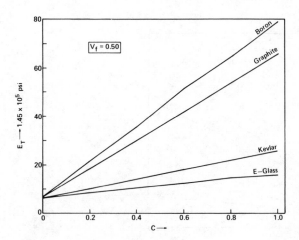

Figure 3.

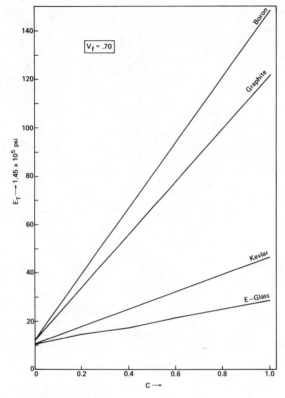

Figure 4.

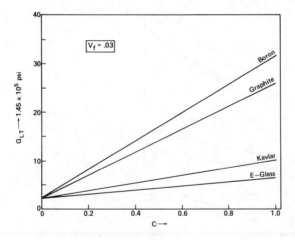

Figure 5.

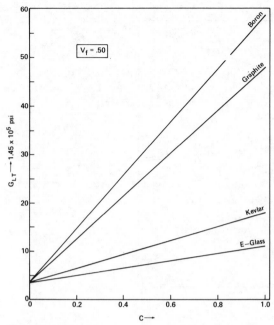

Figure 6.

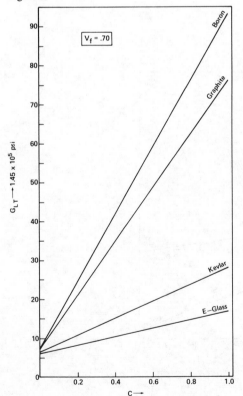

Figure 7.

292

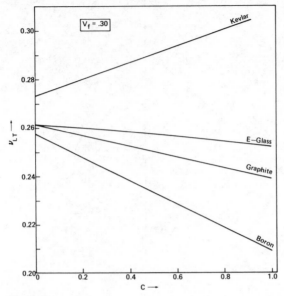

Figure 8.

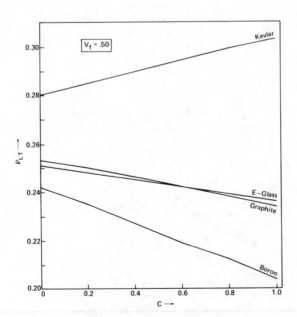

Figure 9.

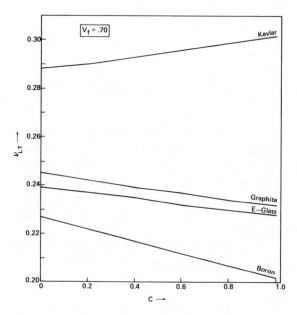

Figure 10.

Empirical equations

While the above equations represent useful formulae, they are not presented in a format which can be easily used by designers. For this reason, some compact and more rapid computational procedure is preferable. To this end the authors of Ref. [23] introduced simplifications in the formulae indicated above which take into account the,
– geometry of the fibers
– fiber packing geometry
– loading conditions
The validity of these simplifications can be measured by comparing the results of the restructured equations with those obtained using the more exact analytical procedures of micromechanics. Thus for predicting the four elastic constants for an orthotropic composite in a plane state of stress, the following equations can be used.

For the unidirectional modulus the rule of mixtures remains a satisfactory computational quantity. That is,

$$E_L = E_f V_f + E_m V_m$$

For the effective elastic moduli E_T or G_{LT}, the following formulae can be used

$$\frac{E_T}{E_m} = \frac{1 + \xi \eta V_f}{1 - \eta V_f}, \qquad \frac{G_{LT}}{G_m} = \frac{1 + \xi \eta V_f}{1 - \eta V_f},$$

$$\eta = \frac{E_{f/E_m} - 1}{E_{f/E_m} + \xi}, \qquad \eta = \frac{G_{f/G_m} - 1}{G_{f/G_m} + \xi}.$$

The factor ξ represents a measure of the constituent element packing geometry and loading conditions. For example, for the transverse modulus E_T, $\xi = 2$ is used while for calculation of the in-plane shear modulus G_{LT} a value of $\xi = 1$ is used. It should be noted that the values of ξ as given above provide reasonable predictions for the elastic constants up to certain volume fractions of fiber packing density and also for reasonable bounds on certain fiber geometries.

For predicting the fourth technical engineering constant, the major Poisson's ratio ν_{LT}, the rule of mixtures can again be used. Thus,

$$\nu_{LT} = \nu_f \nu_f + \nu_m \nu_m$$

When considering anisotropic and twisted fibers, such as yarns, a modification of the above formulae is necessary.

B. Physical properties

An important factor in determining the elastic properties of composites is knowledge concerning the proportion of constituent materials used in the respective lamina/laminates. These proportions can be given in terms of either weight fractions or volume fractions. From an experimental viewpoint, a measure of the weight fractions is easier to obtain than is the corresponding volume fractions of constituent elements. There is however, an analytical connection between these proportioning factors which allows conversion from weight to volume fraction and vice versa. Since volume fractions are key to elastic properties calculations, this connection remains important. The expressions necessary for this development follow.

Definitions

Volume Fraction	Weight Fraction
$V_f = \dfrac{v_f}{v_c}$	$W_f = \dfrac{w_f}{w_c}$
$V_m = \dfrac{v_m}{v_c}$	$W_m = \dfrac{w_m}{w_c}$

f, m, c refer to fiber, matrix, composite respectively

In order to interrelate the above quantities analytically, we make use of familiar density-volume relations. Thus,

$$\rho_c v_c = \rho_f v_f + \rho_m v_m \tag{1}$$

ρ refers to density

The above equation can be rewritten in terms of volume fractions by dividing thru by v_c. Thus,

$$\rho_c = \rho_f V_f + \rho_m V_m$$

Equation (1) can be couched alternately in terms of constituent weights so that,

$$\frac{w_c}{\rho_c} = \frac{w_f}{\rho_f} + \frac{w_m}{\rho_m}$$

Dividing the above equation by w_c we obtain

$$\rho_c = \frac{W_f}{\rho_f} + \frac{W_m}{\rho_m}$$

Introducing now the relationships between weight, volume, and density, we have,

$$W_f = \frac{w_f}{w_c} = \frac{\rho_f v_f}{\rho_c v_c} = \frac{\rho_f}{\rho_c} V_f$$

$$W_m = \frac{W_m}{W_c} = \frac{\rho_m V_m}{\rho_c V_c} = \frac{\rho_m}{\rho_c} V_m$$

The relationship for V_f and V_m in terms of W_f and W_m can now be easily established by inverting the above relations. Further, while the current derivation has been limited but to two constituent elements, the extension to and the inclusion of multiple elements can be easily made.

A relation between weight and volume fractions of fiber or matrix can thus be analytically expressed in terms of the following equations

$$W_f = \frac{(\rho_f/\rho_m)}{(\rho_f/\rho_m)V_f + V_m} V_f \tag{2}$$

$$W_m = \frac{V_m}{\rho_f/\rho_m(1 - V_m) + V_m} \tag{3}$$

296

where,

W_f = Weight fraction of fibers
W_m = weight fraction of matrix
V_f = Volume fraction of fibers
V_m = Volume fraction of matrix
ρ_f = density of fiber
ρ_m = density of matrix
ρ_c = density of composite

Equation (2) has been plotted in Figure 11 for the following fiber types
and corresponding fiber densities as shown in Table 1.

This figure is useful for converting between weight and volume
fraction of fibers for respective materials. Other fiber types, with defined
densities can be added to the plotted data as inferred by interpolation. It
should be noted that for volume fractions of fiber greater than 75% care
in the use of Figure 1 should be exercised. This is due to the fact that
there are theoretical as well as practical limits which can be attached to
the maximum allowable packing densities oriented with different fiber
arrays. As specific examples for the most common arrays encountered the
square and hexagonal, the maximum fiber volume fractions allowed
would be 78% and 91% respectively. These results can be obtained from
simple analysis which is included below for the two most common fiber
packing geometries.

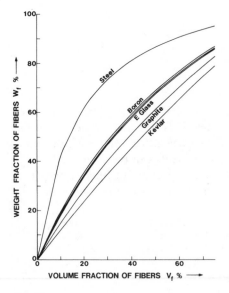

Fig. 11.

Table 1

Fiber Type	Density gm/cm³)
Kevlar 49	$\rho = 1.50$
Graphite	$\rho = 1.90$
E-glass	$\rho = 2.54$
Boron	$\rho = 2.63$
Steel	$\rho = 7.80$

Fiber Packing Geometry

1. Hexagonal Array:

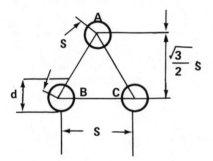

Fig. 12.

Consider triangle ABC

$$A_{\text{total}}(\text{area}) = \tfrac{1}{2}S\left(\sqrt{\tfrac{3}{2}}\,S\right) = \sqrt{\tfrac{3}{4}}\,S^2$$

$$A_{\text{fiber}}(\text{Area occupied by the fibers}) = 3\left(\frac{1}{6}\frac{\pi}{4}d^2\right) = \frac{\pi}{8}d^2$$

Volume Fraction

$$V_f = \frac{A_{\text{fiber}}}{A_{\text{total}}} = \frac{\frac{\pi}{8}d^2}{\sqrt{\frac{3}{4}}\,S^2} = \frac{\pi}{2\sqrt{3}} = 0.907$$

298

2. Square Packing:

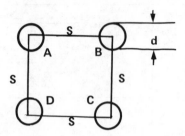

Fig. 13.

Consider a square $ABCE$

$$A_{\text{total}} = S^2$$

$$A_{\text{fiber}} = 4 \times \left(\frac{\pi}{4} d^2 \times \frac{1}{4} \right) = \frac{\pi d^2}{4}$$

$$\therefore V_f = \frac{\pi}{4} = \frac{\pi}{4} = 0.785.$$

References

[1] J.C. Ekvall, "Elastic Properties of Orthotropic Monofilament Laminates," ASME Aviation Conference, Los Angeles, California, 61-AV-56, 1961.
[2] J.C. Ekvall, "Structural Behavior of Monofilament Composites," AIAA/ASME 7th Structures. Structural Dynamics and Materials Conference, Palm Springs, California, 1966, p. 250.
[3] R. Hill, "Theory of Mechanical Properties of Fiber-Strengthened Materials – Self Consistent Model," Journal of Mechanics and Physics of Solids, Vol. 13, 1965, p. 189.
[4] R. Hill, "A Self-consistent Mechanics of Composite Materials," Journal of Mechanics and Physics of Solids, Vol. 13, 1965, p. 213.
[5] J.M. Whitney, "Geometrical Effects of Filament Twist on the Modulus and Strength of Graphite Fiber-Reinforced Composite," Textile Research Journal, September 1966, p. 765.
[6] J.M. Whitney and M.B. Riley, "Elastic Properties of Fiber Reinforced Composite Materials," Journal of AIAA, Vol. 4, 1966, p. 1537.
[7] Z. Hashin, "Assessment of the Self-Consistent Scheme Approximation – Conductivity of Particulate Composites", Journal of Composite Materials, Vol. 2, 1968, p. 284.
[8] Z. Hashin, "On Elastic Behavior of Fiber-Reinforced Materials of Arbitrary Transverse Phase Geometry," Journal of Mechanism and Physics of Solids, Vol. 13, 1965, p. 119.
[9] B. Paul, "Prediction of Elastic Constants of Multiphase Materials," Transactions of the Metallurgy Society of AIME, Vol. 218, 1960, p. 36.

[10] Z. Hashin and W. Rosen, "The Elastic Moduli of Fiber-Reinforced Materials", Journal of Applied Mechanics, Vol. 31, June 1964, p. 223 Errata, Vol. 32, 1965, p. 219.

[11] Z. Hashin and S. Shtrikman, "A Variational Approach to the Theory of the Elastic Behavior of Multiphase Materials," Journal Mechanics and Physics of Solids, 1963, p. 127.

[12] R.A. Schapery, "Thermal Expansion Coefficients of Composite Materials Based on Energy Principle," Journal of Composite Materials, Vol. 2, No. 3, 1968, p. 380.

[13] D.F. Adams and S.W. Tsai, "The Influence of Random Filament Packing on the Transverse Stiffness of Unidirectional Composites," Journal of Composite Materials, Vol. 3, 1969, p. 368.

[14] D.F. Adams and D.R. Doner, "Longitudinal Shear Loading of a Unidirectional Composite," Journal of Composite Materials, Vol. 1, 1967, p. 4.

[15] D.F. Adams and D.R. Doner, "Transverse of Normal Loading of a Unidirectional Composite," Journal of Composite Materials, Vol. 1, 1967, p. 152.

[16] C.H. Chen and S. Cheng, "Mechanical Properties of Fiber-Reinforced Composites," Journal of Composite Materials, Vol. 1, 1967, p. 30.

[17] E. Behrens, "Thermal Conductivity of Composite Materials," Journal of Composite Materials, Vol. 2, 1968, p. 2.

[18] E. Behrens, "Elastic Constants of Filamentary Composites with Rectangular Symmetry", Journal of Acoustical Society of America, Vol. 42, 1967, p. 367.

[19] R.L. Foye, "An Evaluation of Various Engineering Estimates of the Transverse Properties of Unidirectional Composites," SAMPE, Vol. 10, 1966, p. 31.

[20] S.W. Tsai, "Structural Behavior of Composite Materials", NASA CR-71, July, 1964, National Aeronautic and Space Administration CR-71.

[21] J.C. Halpin and S.W. Tsai, "Environmental Factors in Composite Materials Design," AFML-TR-67-423 Air Force Materials Laboratory, Wright-Patterson Air Force Base, Ohio, 1967.

[22] S.W. Tsai, D.F. Adams and D.R. Doner, "Effect of Constituent Material Properties on the Strength of Fiber-Reinforced Composite Materials," AFML-TR-66-190, Air Force Materials Laboratory, 1966.

[23] J.E. Ashton, J.C. Halpin and P.H. Petit, *Primer on Composite Materials: Analysis*, Technonic Publishing Co., Inc., Stamford, Conn., p. 113, 1969.

Appendix 2. Mechanical Property Characterization

As can be discerned from the text material, the role of the engineer in controlling the design process using composite materials requires considerable expertise beyond traditional levels for establishing design criteria. A fundamental input into any design process is the requirement for obtaining the necessary materials properties data as well as establishing the overall material response in order to identify the types of failure events that can occur. Thus the data base for composites is an evolutionary process in which current accepted test standards are being reviewed and revisions adopted as well as composite modes of failure identified and tabulated.

As a ready means of access and awareness to the test procedures in current practice, a test standards has been included. It should be mentioned that in general the engineer executes tests of the following type:

A. *Interrogative*, that is, those examining some aspect, or if seeking fundamental information on certain properties, relations, or physical constants of materials, those using unique test apparatus.

B. *Developmental*, that is, those tests required to obtain additional data to ensure meeting performance specifications on a selected material. In such cases both standard and modified standard test equipment may be used by the engineer.

C. *Standardized*, that is, those tests which utilize controlled test procedures which have been adapted from sanctioned test committee and professional engineering society recommendations. Such tests are almost universally run using commercially available test equipment and with specific geometry specimens.

While all three of the aforementioned type of tests provide important data, it is the standardized test that we tend to rely upon when requiring data for materials. This is especially true since engineers in general wish to be able to duplicate specific tests using accessible equipment rather than designing totally unique test facilities. In view of these statements, the following standards Table 1 is provided for a number of common mechanical tests. Details concerning the test specimen geometry and procedures can be found in the appropriate standard.

Table 1
ASTM test standards

Test Procedure	Filaments (Fibers, Yarns)	Matrices (Thermosets)	Non-Metallic Matrix		Metal Matrix	
			Continuous	Discontinuous	Continuous	Discontinuous
Tension Strength	Textile Filament D2101 Fiber Yarns D2256 Single Filaments D3379	D638	Unidirectional D3090 Off Axis D3039	D638	E 8-61	
Compression Strength		D695	D3410	D695	E 9-61	
Flexure Strength		D790	D790	D790		
Shear Strength		D732				
In Plane Shear			D3518	Rail Shear		
Interlaminar Shear			D2344			
Impact Strength (unnotched)		D256				
Thermal Expansion		D696				

Table 2

Material property	SGlass/XP-251 GPa	psi	EGlass/EP GPa	psi	T300/5208 GPa	psi	AS/3501 GPA	psi	T300/934 GPa	psi	T300/SP-286 GPa	psi	Boron/EP GPa	psi	Kevlar 49/EP GPa	psi
E_x	57.2	8.29×10⁶	60.7	8.8×10⁶	153	22.2×10⁶	138	20.02×10⁶	163.4	23.69×10⁶	151	21.9×10⁶	209 ±6	30.3 ×10⁶	76	11.02×10⁶
E_y	20.1	2.92×10⁶	24.8	3.6×10⁶	10.9	1.58×10⁶	8.96	1.3×10⁶	11.9	1.7×10⁶	10.6	1.53×10⁶	19 ±0.6	2.8×10⁶	5.5	0.798×10⁶
ν_{xy}	0.262		0.23		0.30		0.30		0.30		0.31		0.21		0.34	
G_{xy}	5.9	0.86×10⁶	11.99	1.74×10⁶	5.6	0.81×10⁶	7.1	1.03×10⁶	6.5	0.94×10⁶	6.6	0.96×10⁶	6.4	0.93×10⁶	2.3	0.334×10⁶
$\bar{X}$	1.993	289×10⁶	0.2894	187.0×10³	0.6895	100×10³	1.447	209.9×10³	0.738	107×10³	1.401	203×10³	1.280 ±0.060	185.6×10³	1.400	203.05×10³
$\bar{X}'$	1.172	170×10³	0.8205	119.0×10³	0.7585	110×10³	1.447	209.9×10³	0.724	105×10³	1.132	164×10³	2.500	362.6×10³	0.235	34.08×10³
Y	0.0758611	11×10³	0.04599	6.67×10³	0.0276	4×10³	0.0517	7.5×10³	*	*	0.054	7.8×10³	0.061 ± 0.005	8.8×10³	0.012	1.74×10³
Y'	0.20	29×10³	0.1744	25.3×10³	0.0965	13.9×10³	0.206	29.9×10³	*		0.211	30.6×10³	0.308	44.7×10³	0.053	7.69×10³
S	0.062079	9×10³	0.0448	6.5×10³	0.0621	9×10³	0.093	13.5×10³	0.102	14.8×10³	0.072	10.5×10³	0.105	15.2×10³	0.034	4.93×10³
V_F	0.67		0.72		0.70		0.67		0.60		0.60		0.67		0.60	
Reference	1		2		3		4		5		6		7		4	

Using such tests as described in the standards Table 1, a listing of selected materials properties for continuous filament composites lamina(tes) has been included in Table 2.

The symbols used are:

E_x Modulus of elasticity in the fiber direction

E_y Modulus of elasticity perpendicular to the fiber direction

ν_{xy} Major Poisson's ratio, i.e., $\nu_{xy} = (E_x/E_y)\nu_{yx}$

G_{xy} In-plane shear stiffness

$\overline{X}$ Tensile strength in the fiber direction

$\overline{X}'$ Compressive strength in the fiber direction

Y Tensile strength normal to the fiber direction

Y' Compressive strength normal to the fiber direction

S In-plane shear strength

V_F Fiber volume fraction.

References

1. C.C. Chamis and J.H. Sinclair, "Prediction of Composite Hyral Behavior Mode Simply", SAMPE Quarterly, Vol. 14, 1982, p. 33.
2. Advanced Composite Design Guide, Air Force Materials Laboratory, Wright-Patterson Air Force Base, Third Edition, 1977, Vol 1, p. 1.21-3.
3. E.A. Humphreys, "Development of an Engineering Analysis of Progressive Damage in Composites During Low Velocity Impact", Materials Science Corporation, NASA-CR-165778, 1981.
4. S.W. Tsai and H.T. Hahn, *Introduction to Composite Materials*, Technomic Publishing Co., 1980.
5. Fiberite Corporation, Technical Data Sheet, p. 6.
6. C.C. Chamis (editor), "Test Methods and Design Allowables for Fibrous Composites", ASTM STP 734, 1979.
7. H.T. Hahn, D.G. Hwang, H.C. Cheng, and S.Y. Lo, "Flywheel Materials Technology: Design Data Manual for Composite Materials", Vol. 1, p. 3.4.1.1.

Appendix 3

Answers to selected problems

2.11. $V_f = 24.2\%$, $V_m = 75.8\%$

3.1. 3.168×10^4 lb/in

3.2. $\omega_{11} = 1390$ rad/sec $f_{11} = 221$ Hz.

3.3. $\bar{b} = 78.85$, $\alpha = 1.516$, No, because $|\alpha| \geqslant 1$

4.1. $w(x) = \dfrac{q_0}{48bD_{11}}(2x^4 - 3Lx^3 + L^3x)$

4.2. a. $D_{11} = 1.02 \times 10^4$ in.-lbs.

$$w(x) = \frac{q_0}{24bD_{11}}(x^4 - 2Lx^3 + L^3x)$$

$$w_{max} = \frac{5q_0L^4}{384bD_{11}} = 0.2647''$$

b. $M_{max} = \dfrac{q_0L^2}{8} = M(L/2)$

$\sigma_{max} = \sigma(L/2) = 33,333$ psi

c. 1307 rad/sec = 208 Hz.

d. $P_{cr} = -698$ lbs.

5.3. a. $11.15''$

b. $3.23''$

5.4. a. $\sigma_x(0) = \mp 59,898$ psi

$\sigma_\theta(0) = \mp 1019.$ psi

$\sigma_x(L/2) = 0$

$\sigma_\theta(L/2) = 10,000$ psi.

5.5. $3.32''$

5.6. a. $6.93''$

b. 530.5 psi.

6.1. a. $w_{max} = w(L/2) = \dfrac{q_0L^4}{4\pi^4bD_{11}}$

8.3. $P_{BR} = 2475$ lbs.

Author index

Subject index